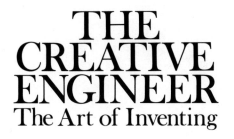

THE CREATIVE ENGINEER
The Art of Inventing

THE CREATIVE ENGINEER
The Art of Inventing

Winston E. Kock

Acting Director
The Herman Schneider Laboratory of Basic
and Applied Science Research
University of Cincinnati
Cincinnati, Ohio

PLENUM PRESS · NEW YORK AND LONDON

Library of Congress Cataloging in Publication Data

Kock, Winston E
   The creative engineer.

   Includes bibliographical references and index.
   1. Inventions—History. 2. Creative ability in technology. I. Title.
T20.K59   600   77-20220
ISBN 0-306-30987-4

© 1978 Plenum Press, New York
A Division of Plenum Publishing Corporation
227 West 17th Street, New York, N.Y. 10011

Printed in the United States of America

PREFACE

The economic growth and strength of a nation are directly related to the ability of its people to make discoveries and their ability to transform these discoveries into useful products. Ninety percent of the increase in output per capita in the United States from 1909 to 1949 has been held to be attributable to technological advances. In this book, we examine the ways in which a number of important new technologies came into being and review the characteristic traits of inventors who *create* new technologies. Ways are suggested that could enable young and old alike to become more creative, and the various benefits they can thereby reap are also discussed.

A high level of creativity is an important asset for a nation, and therefore a knowledge of ways to increase inventiveness can be of great value. University of Cincinnati President Warren Bennis has noted that "creativity is something most of us seem to lose, or let atrophy, as we leave childhood."* To "rediscover it," he continues, "we must find ways of re-creating our sense of wondering why, of heightening, even altering, our consciousness." Thus the earlier in life one seeks to enhance his creativity, the more successful the results are likely to be. We therefore stress, in these discussions, the importance of parents' urging their children to become creative, so that their lives will be made fuller, happier, and more rewarding. But the process of influencing youngsters is not an easy one, because the bright, up-and-coming young person demands that he be shown by many examples that his parents' advice is sound.

* Warren Bennis, *University of Cincinnati Horizons*, p. 21 (January 1974).

In this connection, one of my long-time friends, Nobel Laureate Luis Alvarez, pointed out to me recently that the majority of books on creativity have been written by authors who have not demonstrated extensive creative abilities. Because Alvarez has, over the years, been aware of my numerous inventions, he encouraged me to describe in great detail my various experiences in the fields of invention and creativity. In this book, I have followed this advice, because I belive that if youngsters can be shown concrete examples of how inventions have led to *rewards* the task of convincing them becomes a much easier one. (Chapter 9 deals with rewards at some length.)

Most of the references and photos dealing with my experiences have been relegated to the Appendix, enabling the reader to skip over them if he desires. But because they are there, parents can use them in helping their children to see the advantages of being creative. For innovators do reap benefits, and they also benefit their country and the world. The technological advances described were selected partly because I was involved in, or in close contact with, their progress, but more importantly because they exhibit certain attributes of invention that can suggest how to develop one's latent creativity. They were also chosen because they have largely involved engineers having interdisciplinary experience; it has been predicted that the demand will be strong in the 1980s for graduates who can apply engineering principles to medical, biological, and other sciences.

At Fairleigh Dickinson University, Professor Harold Rothbart has instituted a course which he refers to as "an invention workshop, intended to relate inventive activities of all sorts for the purpose of increasing the potential for creating original products." Students are encouraged, in a free environment, to evoke a flow of unstifled, creative expressions of ideas. "Everyone should share the excitement of invention," he continues, "and this interdisciplinary activity is catalytic to other creative activity."* The value of the interdisciplinary activity that Rothbart mentions was stressed in a recent report of the Enrico Fermi Institute at the University of Chicago,† a laboratory noted for the work done there during the early stages of the U.S. nuclear program. The report notes that the "interdisciplinary character of the Institute has strengthened the traditional departments of instruction within the

---

* *Industrial Research,* p. 80 (November 1974).

† *Enrico Fermi Institute* (University of Chicago pamphlet distributed to members of the Physical Sciences Visiting Committee, November 1, 1974).

University by providing a common base of intellectual inquiry and the free exchange of ideas. Out of this interdisciplinary environment have come fresh ideas leading to new fields of research." Through my work at the Herman Schneider Laboratory at the University of Cincinnati (which concentrates its support on interdisciplinary research programs), I have observed at first hand that this need to involve two or more disciplines has led to numerous proposals being submitted which describe novel and interesting concepts. Even a limited knowledge of several disciplines can often enable the owner of this knowledge to contribute in an important way to these disciplines. The existence of a new development in one field permits the interdisciplinary mind to consider related applications of this development in other fields, often with fruitful consequences. Obviously, a development which is novel in one field will constitute a novel development in another field when successfully applied thereto.

The level of the discussions here has been set so as to make them understandable to a second- or third-year college student, so that he can consider early in his college career the desirability of selecting courses on subjects in fields other than his major one. The examples which follow are intended to encourage the reader to acquire a multidisciplinary interest, thereby permitting him to add to the world's store of scientific knowledge and its application. I also hope that my background of patent activity (235 issued U.S. and foreign patents stemming from my 14 years as a member of the technical staff at the Bell Telephone Laboratories, plus many others at the D. H. Baldwin Company, the Bendix Corporation, and NASA), will enable me to communicate ways which can help the reader experience the excitement stemming from discovery and invention. Because of the level, the book could well find use as auxiliary reading in the sophomore and junior years.

For various photos and figures, I wish to express my indebtedness to the D. H. Baldwin Company, the Bendix Corporation, the University of Michigan, the National Aeronautics and Space Administration, the Communications Satellite Corporation, the Bell Telephone Laboratories, and the American Telephone and Telegraph Company.

Winston E. Kock

Ann Arbor

CONTENTS

CHAPTER 4
LASERS

CHAPTER 5
TRAITS

CHAPTER 6
WAVEGUIDES

CHAPTER 7

LENSES

CHAPTER 8

COMMUNICATIONS SATELLITES

CHAPTER 9
WHY INVENT?

CHAPTER 10
RADAR

CHAPTER 11
HOLOGRAPHY

CHAPTER 12
PICTUREPHONE

STIMULATING ONE'S OWN CREATIVITY

The motivation for creativity can be simply the intriguing pleasure of seeing one's own ideas materialize into something worthwhile. We can easily imagine the many such pleasures experienced by Leonardo da Vinci, Galileo, Hertz, and similarly by Bardeen and Shockley with their beautiful semiconductor theory, by Brattain with his clever idea of the two-point-contact transistor, and by Gabor through his simple idea of recording the holographic interference pattern rather than the photographic light pattern. None of these innovators concerned himself with how his ideas might affect the world. They all generated *ideas* and would still have been very happy if the only result had been a successful experiment.

Ideas, even associative ideas (those that transfer a new concept from one field to another), are what brings the world new technologies. Michael Wolff* describes how the snap-action disk was invented by the MIT student Albert Spencer during a nighttime summer job tending a sawmill furnace in Maine. Spencer became intrigued with the fact that the sheet metal door *snapped* when the fire got very low. In his *wondering why,* the snapping set him to thinking, and soon he had hammered out a can of bimetal that would snap hard enough to jump off a table. By the 1940s, a multimillion-dollar industry had grown up on the basis of the "Spencer disk."

So what are some of the ways for extending our own "idea con-

---

* Michael E. Wolff, "Inventing at breakfast," *IEEE Spectrum,* pp. 44–49 (May 1975).

sciousness," our own inventiveness, ways for rediscovering and making use of that creative talent we all possess to some degree as children? For, as Warren Bennis reports, "an artist who worked with slum children, letting them draw or paint whatever their minds suggested, concluded that *every* child under ten can create things that have the unmistakable air of distinctive originality."* He suggests that "the reason for this is that, to every child, the world around him is a totally new discovery; the green grass, the nodding trees, the grace of animals, the poetry of wind, the grave silence of snow, the re-creating sun—all, each day, are born over and over. The child encounters this miracle with a sense of wonder, one that his elders lose as the familiarity, or tedium, of daily life shuts them out. In other words, creativity is something we all have, yet manage to lose."

How can we rekindle this sense of wonder, this sense of *wondering why,* so that more of us can share the excitement of invention? One way, suggested by Harold Rothbart, is to "encourage the flow of unstifled, creative expressions of ideas."

Along another line, Bennis observes that as we leave childhood "we do not really see the world around us, we see only a *'gloss'* of stereotyped expectations. We may see a leaf, but not the magic, the harmony, the incredible order of the intricate veining of the leaf." He mentions a Yaqui Indian wizard who taught a visitor that he could discover a "separate reality" quite different from our ordinary reality. The secret was learning to "'stop the world'—to break through the *gloss* by which we see only what we expect to see." He goes on to note that Eastern mystics have shown that periods of meditation can heighten one's consciousness, suggesting that all of us could benefit from such "private cathedrals of contemplation."

Along these lines, Bennis recommends that individuals seeking to reclaim their creativity must first break the pattern of the familiar, for instance, by taking up some new interests, redeveloping atrophied talents, or rekindling old enthusiasms. He points out that "the more our work makes us *specialists,* the more we must strive to become *generalists.*"†

The majority of the following chapters describe inventions dreamed up or contrived by *nonspecialists,* by persons who had a general

---

* Warren Bennis, *University of Cincinnati Horizons,* pp. 21–22 (January 1974).
† Bennis, *op. cit.*

knowledge, even though somewhat limited, of fields other than the particular field of science involved in their invention or discovery. It is this nonspecialist nature which led us to stress, in the Preface, the value of multidisciplinary knowledge. Perhaps even more important, however, are the pervading characteristics of inventors, their constant feeling of curiosity, their continual search for new possibilities, and their refusal to accept stock answers. For example, once there was a young boy whom his neighbors (who were annoyed by his constant questions) jokingly called "Mr. Wonder Why." Everyone scoffed at him until suddenly he began making unusual discoveries, and then his detractors had to backpedal vigorously. The constant curiosity of "Mr. Wonder Why" is a characteristic found in almost all innovators, and they are not deterred by the scoffing of others. Charles F. Kettering had his own argument for taking the edge off such derision. He said, "People think of the inventor as a screwball, but no one asks the inventor what he thinks of other people."

The famous MIT research engineer H. Stark Draper (who gave his name to the MIT Draper Laboratory) also had some favorite sayings about research. Here are a few: "Research is a gamble. It cannot be conducted according to the rules of efficiency engineering. The best advice is, don't quit easily. Don't trust anyone's judgment but your own. The best person to decide what research work should be done is the man who is doing the research."

Draper's admonition about not quitting easily is reminiscent of one of Will Rogers's well-known sayings, "Even if you're on the right track, you'll get run over if you just sit there." And Draper's contention about efficiency engineering was echoed recently by E. Bruce Peters,* who noted that creative individuals are not likely to fit the common mold of organizational activities but must be granted far greater autonomy and independence than other employees, because the creative person is driven by his work, rather than by directives from his boss. Peters also contends that creative individuals cannot be forced to produce usable ideas on a regular, measured basis.

Next, let us turn to a discussion of *who* can acquire, or reacquire from his childhood days, a stronger talent for creativity. First, must the highly creative person have received unusually high grades in school?

---

* E. Bruce Peters, "Creativity in Conflict," *Industrial Research,* pp. 69–82 (April 1975).

Roland S. Illingsworth has compiled a list of "school failures"* which includes "bottom of the class," Thomas A. Edison; "poor mathematicians," Benjamin Franklin; "expelled from school," Wilhelm Roentgen; and "mentally slow," Albert Einstein.

Second, must the person seeking to improve his creative ability be young? Apparently not. Lissy F. Jarvik† has shown that the common belief that intellectual functioning peaks at about age 17 and then declines progressively is not based on fact; his studies show that a decline in knowledge or reasoning power does not generally occur even during one's 60s and 70s.

Finally, has a hard scientific analysis of creativity been made? Here the answer is suggested by a *Science News* report‡ of the results of tests made by Colin Martindale, who studied at some length the brain-wave patterns (electroencephalograms) of high-, medium-, and low-creativity people and discovered interesting facts about their "cortical arousal":

> when incoming information is on its way to the brain, the cortex is alerted. During sleep, with little input, cortical arousal is at a minimum. It increases as people go from sleep to states of reverie and daydreaming to alert concentration and finally to emotional agitation and panic. With electroencephalograms (EEG's) the degree of cortical arousal can be measured. The slow alpha waves, for instance, are produced during periods of complete relaxation and meditation. As arousal increases, alpha decreases. And when people react to strong stimuli, for instance, alpha is blocked and replaced by faster waves.
>
> Martindale's studies of the EEG's of high, medium, and low creative people suggest that cortical arousal is directly related to creativity. Brain-wave measurements were taken during a resting state. Highly creative people were found to produce alpha waves only 37 percent of the time (62 percent of the time they produced the faster waves). Medium and low creative people produce alpha up to 50 percent of the time. In other words, when tested during a resting state, creative people have higher levels of brain-wave activity than average people. Highly creative people also had higher levels of skin conductance, another measure of arousal.
>
> Martindale's findings suggest that creative people may be more sensitive to and conscious of incoming stimuli. Experiments designed to test the sensitivity of individuals to various stimuli did show that creative people

---

* Roland S. Illingsworth, in *Learning, the Magazine for Creative Thinking,* as reported in *Reader's Digest,* p. 102 (September 1975).

† Lissy F. Jarvik, UCLA, as reported in the *Reader's Digest* (September 1975).

‡ Colin Martindale, University of Maine, as reported in "Creativity and cortical arousal," *Science News, 108,* p. 53 (July 26, 1973).

tend to amplify sights, sounds, and textures. Because they are oversensitive, it seems that creative people block out a great deal of alpha activity.

There are at least some indications, thus, that childhood creativity can be reacquired by adults. By widening our interests, by increasing our sensitivity to incoming stimuli, and by not suppressing thoughts of wonder and the desire to know why, we can make our minds more alert to new ideas and new discoveries.

Benjamin Franklin said, "The used key is always shiny." To keep one's research talent shiny, it, too, requires use, and to keep it in use requires challenging areas of exercise. The provocative urgency of World War II triggered a productivity of great distinction, all the way from radar through jet aircraft and nuclear technology to rocketry. Most of today's research programs can also provide a stimulating and inspiring atmosphere, and this atmosphere should ensure that the brilliant past sparkle of research-conscious minds will, like Ben Franklin's key, continue to remain bright.

In summary, we see that some of the methods for enhancing our own creative ways and thoughts appear to be the following: extending our interests into new fields, acquiring a more inquisitive, "wonder why" trait about subjects and concepts we now take for granted, and spending more time contemplating and meditating about our own ideas and about odd, newly reported discoveries and theories. These and other exercises of the mind should increase our cortical arousal and get us closer to the thought habits of those who have demonstrated an inventive, innovative nature.

We shall examine further these *traits* of inventors in a later chapter, and also, in another chapter, some of the *reasons* for their innovative actions. But first let us look at some aspects of an exceedingly important discovery, the invention of the transistor, and then review how some other new fields of technology came into being.

2

THE TRANSISTOR

EARLY SEMICONDUCTOR RESEARCH

During the early part of World War II, an important electronics development was in its early stages. Scientists the world over were beginning to understand more and more about the inner workings of matter, about its electrons and protons and how their motion within various materials could be controlled. One branch of this new electronics development benefited very significantly from contributions made by Professor K. Lark-Horovitz at Purdue University.

Electric currents travel in conductors (such as in the metal copper), and they cannot travel in insulators or nonconductors (such as air). Professor Lark-Horovitz and his associates were then analyzing, both theoretically and experimentally, a different kind of substance, one called a "semiconductor," so named because it is neither a good conductor nor a good insulator. They were finding that many such newly available substances worked extremely well as rectifiers, two-"electrode" devices which can, for example, be used to change alternating current (such as the 110 volt house current) to direct current (such as the battery current in automobiles). One particularly promising semiconductor material was found to be the element germanium.*

Shortly after the war, in 1946, the General Assembly of the Union Radio Scientifique Internationale (URSI), a very erudite international

* K. Lark-Horovitz, A. E. Middleton, E. P. Miller, and I. Walerstein, "Electrical properties of germanium alloys," *Phys. Rev., 69,* p. 258 (March 1946). V. A. Johnson and K. Lark-Horovitz, "Theory of thermoelectric power in germanium," *Phys. Rev., 69,* p. 259 (March 1946).

radio and electronics body, was held in Paris. Most of the 90 papers presented there reported on developments in the radio field, with authors including British scientists Sir Edward V. Appleton, E. C. S. Megaw, H. G. Booker, and F. Adcock, French scientist Y. Rocard, Australian scientist E. C. Bowen, and U.S. scientists H. LeCaine, Newbern Smith, R. M. Fano, and W. G. Tuller. One Bell Laboratories paper presented there, number 64 (see Appendix 1), on microwave measurements* was prepared by C. C. Cutler, A. P. King, and myself. Another paper, number 51, presented by Lark-Horovitz, was entitled "Electrical Semi-conductors and Their Uses." This paper, describing his pioneering work, heralded important developments.

THE BELL LABORATORIES PROGRAM

During the postwar period, activity in the semiconductor field increased markedly in many scientific laboratories throughout the world. But of greatest importance in this growing activity (based on its later results) was a research program initiated by the U.S. scientist Dr. Mervin J. Kelly, then Executive Vice-President of the Bell Telephone Laboratories. This program involved the gathering together of a sizable and very impressive group of theoretical and experimental scientists to do research on semiconductors (such as germanium). When this program was announced, many were surprised that the Bell Telephone Laboratories would embark on a major technological project simply in the *hope* that it would yield useful, practical results. This new semiconductor group was located at the then quite new Murray Hill Laboratories of Bell, and the scientists and engineers involved in it were widely referred to, even within the Laboratories, as the "ivory tower" group. A typical question was "In what possible way can a group like that provide assistance to engineers engaged in the Bell Telephone System's communications problems?"

With relative unconcern over such attitudes, the ivory tower group began at once to make impressive strides, and soon an elegant theoretical analysis of the behavior of semiconductors resulted, called "Surface States." On the basis of this analysis, ivory towerites John

---

* C. C. Cutler, A. P. King, and W. E. Kock, "Microwave antenna measurements," *Proc. IRE, 35,* p. 1462 (1947).

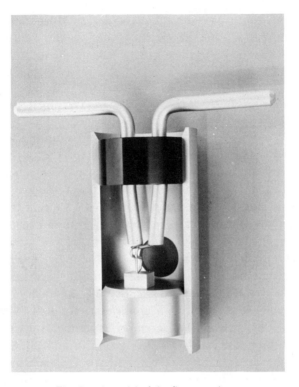

Fig. 1.   A model of the first transistor.

Bardeen, William Shockley, and Walter Brattain came up with an experiment that proved to be world shaking.* The device involved in the experiment was constructed by Brattain (he was the experimentalist of the three) and consisted of two sharply pointed wires pressing down on a piece of germanium having a connection made to its bottom side (Fig. 1). Just as Dr. Lee DeForest's first three-electrode radio tube device had made history, so this three-electrode semiconductor device was also destined to make history, for it could do what the earlier two-electrode semiconductor devices (rectifiers) could not—it could *amplify*

---

* W. Shockley and G. L. Pearson, "Modulation of conductance of thin films of semiconductors by surface charges," *Phys. Rev., 74,* p. 232 (1948). J. Bardeen and W. H. Brattain "The transistor, a semi-conductor triode," *Phys. Rev., 74,* p. 230 (1948). W. Shockley, "Holes and electrons," Nineteenth Joseph Henry Lecture delivered before the Philosophical Society.

electrical signals, just as radio tubes could. The world's first *transistor* had been constructed, and Dr. Kelly's ivory tower group had brought about a revolution in electronics.

PREPARATIONS FOR THE TRANSISTOR ANNOUNCEMENT

Those first successful experiments involving the two-point-contact device occurred late in December 1947, with an official "witnessing" of its amplifying capabilities taking place on the afternoon of December 23.* Scientists and executives at Bell knew at once that the innovation was a momentous one, and a plan was drawn up to provide the public with ample information about the discovery and yet ensure maximum future benefits to Bell System customers. The arrangement included a delay in the announcement to the press and a gathering together of several dozen Bell Laboratories scientists of varied disciplines, all of whom had demonstrated, through applications for Bell System patents, a capability for invention. The December experiments were described to the group (named the "surface states group"), and all were urged to ponder ways for applying the new amplifier in various fields. They were also instructed to conduct experiments to verify their concepts, and to enter the concepts and have them witnessed in their Laboratories notebooks. Patents could then be applied for on the more important entries, thereby avoiding, to a large extent, the possibilities of non-Bell corporations later patenting such concepts with the Bell System then having to pay royalties on those patents.

I consider myself most fortunate to have been chosen a member of that surface states group; in all probability, my selection was based primarily on my earlier patent record. This included the issuance of 29 U.S. and foreign patents prior to my joining, in the summer of 1942, the Bell Laboratories, and over 100 Bell patent applications (U.S. and foreign) made between 1942 and 1948. Immediately following my selection, my acoustics research subdepartment began, as did others in the surface states group, devising new concepts and experimentally testing them, with many of these then being judged valid candidates for patent action. A 6-month time limit loomed, because June 30 was selected as

* *Electronic News,* p. 51 (December 18, 1972).

the day on which Dr. Ralph Bown, Vice-President for Research at Bell
Laboratories, would make the public announcement.

PATENT PRECAUTIONS

The preparing of patent applications takes time, and—even before
the transistor was devised—Dr. Bown had sent a letter to all Research
Department members urging that, because the Laboratories Patent
Department was overloaded with wartime inventions and hence facing
serious delays, certain procedures be instituted to speed up the patent
application process. Because many of the innovations that the surface
states group came up with led to striking demonstration items for the
press show, I sent a memorandum to Dr. Bown which included cau-
tionary sections such as "this device uses a cathode follower which is not
yet protected; the circuit details should be hidden from view" and "the
audio oscillator circuit may have patentable novelty; only the transistor
should be shown."

The Patent Department also recognized the problem and sent Dr.
Bown a memorandum of its own. Its opening sentence was "This is to
comment from a patent standpoint on the proposal to disclose and
demonstrate certain Surface States equipment to the press on June 30,
1948 as outlined in a memorandum from Mr. W. E. Kock to Mr. R.
Bown dated June 8, 1948." It continued: "To fulfill our obligation to
the Western Electric Company [the manufacturing arm of the Bell
System] with respect to possible inventions that Western may wish to
patent in foreign countries, it will be necessary to circumscribe carefully
the disclosure that is made public on June 30, 1948." A later paragraph
opened with "The following specific comments supplement those set
forth in Mr. Kock's memorandum." The remainder of the letter
specified other demonstration items for which public disclosure at the
early date of June 30 could be harmful to full patent protection.

THE TRANSISTOR ANNOUNCEMENT

Great care was taken and great effort was made, both by the
engineers and scientists involved and by the Patent Department, to make
certain that the press announcement would not only provide the public

Fig. 2. Bell Laboratories Vice-President Dr. Ralph Bown describing a 100-to-1 scale model during the first announcement to the press of the discovery of the transistor.

with ample coverage of the invention but also protect the Bell System and its customers. As M. J. Kelly, who became President of the Laboratories in 1951, noted in 1953 in an article entitled "The First Five Years of the Transistor"*: "It is interesting to recall the words used by Laboratories Vice-President Ralph Bown to tell about the invention. He said in part: 'What we have to show you today represents a fine example of brilliant individual contributions growing out of basic research in an industrial group framework . . . . A considerable number of people have been working hard on this matter to bring it to the stage you will see it today. Physicists, chemists, metallurgists, engineers, laboratory and shop technicians, auxiliary and office personnel—yes, even executives—have played a part.'"

It is not hard to envision how pleased all of those persons mentioned by Dr. Bown were by his public recognition of their efforts, and the deep satisfaction they experienced through this bestowing of due credit. The acknowledgment provided a strong incentive for others; in fact, the entire sequence of events constitutes a beautiful example of how the inventive spirit of individuals and groups can be strengthened.

* *Bell Telephone Magazine, 31,* p. 73 (1953).

Figure 2 shows Ralph Bown describing the transistor at the June 30 press show, using a 100-to-1 scale model.

Unfortunately, as was noted by Dr. Kelly, "the announcement received little public notice—one of the most restrained send-offs in recent memory, according to *Fortune* Magazine." The *New York Times* writeup of it was limited to a half-dozen lines on the last page in the Radio Program section. Accordingly, to more fully inform others in the Laboratories of the historic development, and also to imbue them with a sense of camaraderie with the group of innovators, the Publications Department arranged for the New York press demonstration to be repeated several times at the Murray Hill Auditorium for Laboratories members. Called on to substitute for Vice-President Bown at many of these presentations, I noted that it was a "very difficult spot to fill" and continued, "To ensure that nothing be omitted from this talk, I shall use Dr. Bown's original notes," adding, however, a postscript naming "those who helped make this show possible."

THE BENEFITS

To conclude this discussion of the first (point-contact) transistor, some excerpts from my talk before the New Jersey Press Association at the Murray Hill Auditorium help to accentuate how such an invention can advance the world's technology in many fields. These advances, which we now take for granted, gave birth to huge industries and provided for a continually increasing standard of living. Here are the excerpts:

> In describing the transistor to you, the most significant thing I can say about it is that it is an electrical amplifier, a means for building up the strength or loudness of an electrical signal. This is significant because in the whole vast structure of electronics which has sprung up since the turn of the century, there has been only one other important electrical amplifier, the vacuum tube, exemplified here by this typical radio tube of your home receiver. Furthermore, it is just this amplifying property of the vacuum tube that has made possible all of present-day electronics. Without an amplifier, we would not now have long-distance telephone, we would not have radio, radar, television, or any of the electronic advances we now take for granted. In fact, it was the early endeavors of the Bell System to provide cross-country telephony which led Bell researchers to perfect the vacuum tube amplifier invented by Lee DeForest.

So here is a discovery, the transistor, which challenges for the first time the vacuum tube, the heart of all electronic devices. It challenges it because it is simpler to make, it has no hot filament to light up and to burn out, and it needs no glass bulb to hold a vacuum.

This large object, which looks something like a hot water tank, is in reality just what a transistor would look like if blown up a million to one in volume. It is cut away on this side to show its simple construction. Just two fine wires pressing against a small piece of semiconductor, in this case the metal germanium. Yet with appropriate battery conditions, a signal coming in on this lead will appear at this other lead strengthened or amplified 100 times.

This signal might be your own voice in a long-distance telephone conversation, where the length of telephone line is so great that the sound would be too feeble at the other end without a booster amplifier of this sort. Just as the sound of an overhead airplane becomes weaker as the plane moves away, so your voice signal becomes weaker as it spans the country along the telephone lines and cables, and many separate vacuum tube amplifiers are now used to boost the signal back to its original loudness. So you see this tiny transistor, which can amplify a signal 100 times, three in tandem therefore amplifying a million times, is of great importance in telephony.

To demonstrate this property of amplification, we have here a transistor oscillator and amplifier. Oscillators are used to a great extent in telephone work for signaling and other purposes. Here a transistor oscillator generates the tone you now hear, and by switching in the second transistor, connected as an amplifier, you notice the tone becomes louder. Just as this tone was made louder, so your voice would become louder in a telephone circuit by the use of the transistor.

The amplifying property of the transistor is not limited, however, to speech signals. This tone you hear is approximately a 1000-cycle tone and telephone speech involves signals up to about 3000 cycles. But television pictures, which the Bell System is also called upon to pipe all over the country for network television, involve signals extending up to several billion cycles. And we can now show you that the transistor takes even these wide bands in stride [television amplifier demonstrated].

The television band, however, finds the transistor at its limit in frequency, and because it is small, it cannot at present handle large powers. But within these two limitations, it will do just about everything a vacuum tube will do, so that many electronic devices can now dispense with vacuum tubes and use the simpler transistor instead. We have shown you some of its applications in electronic equipment used in the telephone system, but perhaps the electronic device with which you are most familiar is the home radio set.

Every home radio set made today has at least three or four radio tubes and the larger ones have ten or more. Before the advent of the transistor,

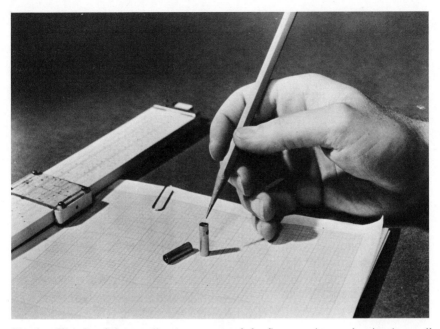

Fig. 3.   The tip of the pencil points to one of the first transistors, showing its small size.

there was no conceivable way of building such a radio set without vacuum tubes, as there simply was no other amplifier available. But here we now have a radio set without a single vacuum tube in it, all of the amplification being accomplished by transistors [radio set demonstrated].

Well, just as this radio set without radio tubes is perhaps startling to you, so the possibilities of the transistor in the Bell System, where millions of vacuum tubes are in continuous service, appear startling to telephone engineers. And to us in the Bell Laboratories it is just another milestone along the road toward our constant goal of better telephone service at cheaper cost.

The tiny size of an actual transistor (Fig. 3) as compared to that of a vacuum tube constituted one of its most important advantages. In 1956, Bardeen, Shockley, and Brattain were awarded the Nobel Prize in Physics for their invention of the transistor. Figure 4 shows their picture, taken in 1948, gracing the cover of the 1956 Bell journal (the *Bell Laboratories Record*) which featured the Nobel Prize award.

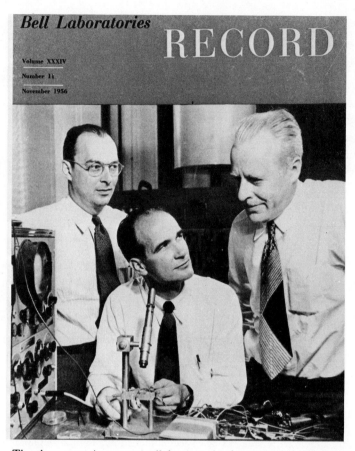

Fig. 4. The three most important collaborators in the research leading to the transistor. Left to right, Dr. John Bardeen, Dr. William Shockley, and Dr. Walter Brattain. In 1956, they received the Nobel Prize for their work.

FROM SURFACE STATES TO SOLID STATE

Let us move on to an extension of this example of creativity in the semiconductor picture. The "surface states" concept, suggesting that it was the electronic phenomena occurring at the *surface* of the germanium which were responsible for the properties observed, was indeed very valuable, for it led to Walter Brattain's construction of the first successful transistor. And, because this first model would amplify only when the surface distance between the two point contacts (called the "emitter" and "collector" contacts) was quite short (as seen in Fig. 1), this existence of a minimum distance appeared to constitute convincing proof that the transistor effect was indeed due to a surface phenomenon.

Here is where a second case of "wonder why" arose. Another Bell Laboratories semiconductor scientist, John Shive, curious as to whether the electric currents in the transistor were on the surface, decided to try new arrangements for positioning the two point contacts. In the course of this investigation, he discovered that amplification could also be obtained when the emitter and collector points were placed on opposite sides of a germanium wedge. In his structure, the germanium wedge was narrowed down to a sharp edge, and the point contacts were placed on opposite sides, at a point where the wedge was only a few thousandths of an inch thick.

The success of the wedge device seemed to indicate that the current passing between emitter and collector points might actually be passing *through* the semiconductor and not around the wedge surface. Hence, if the effect were *not* due to a surface phenomenon, it must have been caused by an amplification process occurring within the semiconductor itself. Obviously, this conclusion would generate important theoretical repercussions in regard to the then popular surface states hypothesis. However, because the points in the Shive wedge were rather close to the sharp end of the wedge, one could not be truly certain that surface effects were not responsible for the amplifying action.

It was then that I, along with my Bell Laboratories colleague R. L. Wallace, Jr., extended John Shive's innovation by devising a way to ascertain definitely whether the effect was due to a surface effect or due to a phenomenon occurring within the *solid* portion of the semiconductor (a prelude to the now widely used description for transistor-related technology, "solid state"). The idea was to fabricate a germanium *disk* rather than a wedge, having a dish-shaped (spherical) depression on each of its two faces. Point contacts would then be pressed against the central portions of these two depressions, where the thickness would be comparable to the few thousandths of an inch used by Shive in his wedge.

THE DOUBLE-DIMPLE TRANSISTOR

Needless to say, the hollowed-out, dimpled germanium disk *was* fabricated, it did amplify, and a paper on it was presented on January

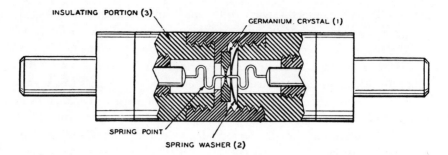

INSULATING PORTION (3)

GERMANIUM CRYSTAL (1)

SPRING POINT

SPRING WASHER (2)

Fig. 5. A cross-sectional view of the "double-dimple" or coaxial transistor.

31, 1949, at the AIEE meeting in New York City (see Appendix 2). A cross-section of the first experimental model is shown in Fig. 5. The unit started out understandably with the name "double-dimple" transistor, but as can be seen in Fig. 5 the connection to the body of the germanium (the "base" connection) was made over the entire rim of the disk, thereby permitting a cylindrical shield to surround the entire device. As the paper describing this unit stated: "The germanium disk, normally grounded electrically, is seen to provide an electrostatic shield between the emitter and collector points, and all three parts—emitter point, collector point, and germanium disk—are seen to be coaxial."* Accordingly, the name "coaxial transistor" was chosen as the official name of this *solid-state* amplifier. Another view of this unit is shown in Fig. 6. A photo of the author and John Shive with R. L. Wallace holding a model of it is shown in Fig. 7. This is a photo of the cover of the official journal of the American Institute of Electrical Engineers.

The wedge and the double-dimple transistor experiments led to a significant further understanding of semiconductors. The coaxial construction was successfully applied by Shive to a phototransistor (Fig. 8) and by Wallace and me to a form of amplifying microphone (Fig. 9). But of greatest importance was the recognition of the solid-state character of the transistor amplifier phenomenon. This recognition soon led William Shockley and his colleagues at Bell Laboratories, Morgan Sparks and Gordon Teal, to announce the first truly solid-state device, the junction transistor.† It substituted solid metallic junctions for the

* W. E. Kock and R. L. Wallace, Jr., "The coaxial transistor," *Electrical Engineering, 68,* pp. 222–223 (March 1949).
† W. Shockley, M. Sparks, and G. K. Teal, "The *p-n* junction transistor," *Phys. Rev., 83,* pp. 151–162 (July 1951).

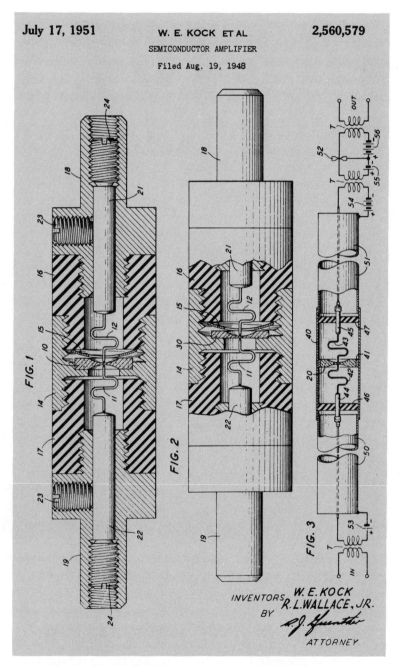

Fig. 6. The Kock–Wallace patent on the coaxial transistor.

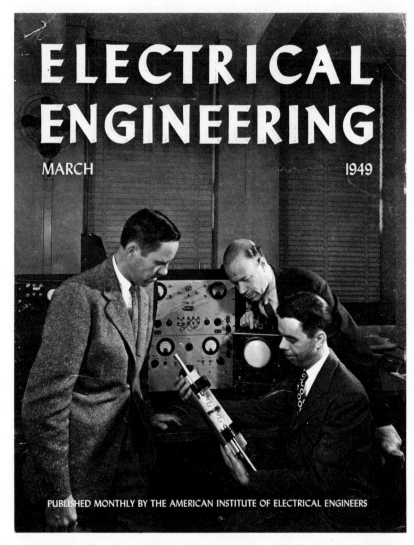

Fig. 7.   Bell scientist R. L. Wallace, right, holds a greatly enlarged model of the coaxial transistor. J. N. Shive is behind Wallace and the author is at the left.

relatively low-reliability point contacts of the first transistors, and this form is the one now used universally. It soon was recognized that these newer solid-state devices were capable of demonstrating a tremendous improvement in reliability over the earlier accepted standard in the field of electronics, the vacuum tube (and also over the early point-contact

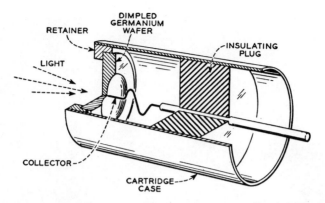

Fig. 8. One of the "spinoffs" of the dimpled transistor was a dimpled "phototransistor" which provided amplification of the photoelectric signal.

transistors). This high reliability is largely responsible for recent successes in long-time operation of electronic devices without attention or repair, such as in computers, in the man-on-the-moon space program, and in the long-lifetime communications (television) satellites (where repair service would be rather awkward!). The "solid-state design" was

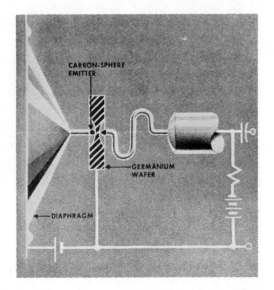

Fig. 9. A double-dimple microphone, wherein the varying pressure (due to the sound waves impinging on the diaphragm at the left) caused the current in the emitter to vary. Transistor amplification then occurred, resulting in a "high-output" microphone.

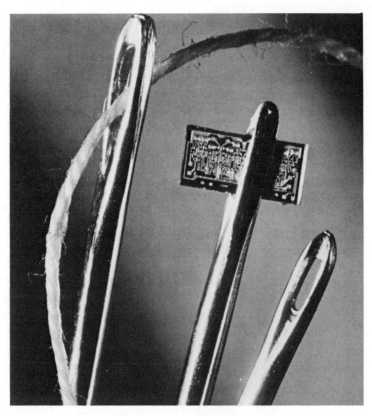

Fig. 10.   A demonstration of how tiny some of the solid circuits are that are used today. Courtesy Eastman Kodak.

eventually extended to many other component devices, such as registers and capacitors, and 20 years after the first transistor, complete electronic circuits, comprising tens of components, could be made small enough to literally pass through the eye of a needle (Fig. 10).

TRANSISTOR APPLICATIONS

Because of the tremendous growth in the use of transistors in computers in recent years, it is of interest to note that the 792-page book entitled *The Transistor* prepared by Bell Laboratories in 1951 (which, incidentally, also contained the paper referred to earlier by R. L.

Wallace and me on the coaxial transistor) included numerous papers which discussed devices useful in computers. These included binary counters, shift registers, encoders, and anticoincidence gates.

Ivory tower creativity had been demonstrated, and the transistor along with the germanium wedge and double-dimple experiments led to advances in electronics which were to prove valuable to the entire world. The *reliability* that solid-state devices provided can be said to have been one of the most important factors in the success of many outstanding new programs—including the space effort, which, as we shall see later, made possible communications satellites. The transistor gave the world not only transistor radios and transistor television sets but also transistorized computers, electronic organs, transistor door chimes, transistorized fuel injection systems to make automobiles more pollution free, and lightweight, high-fidelity amplifiers for music.

LEAVING THE BEATEN TRACK

There is a statue of Alexander Graham Bell in the lobby of the Murray Hill Laboratories, and on this statue can be found some very valuable advice relating to invention and creativity: "Leave the beaten track occasionally and dive into the woods. You will be certain to find something you have never seen before." The invention of the transistor exemplifies well Alexander Graham Bell's exhortation, and it also has proved correct Dr. Kelly's conviction that a dive into the "semiconductor woods" by a communications engineering groups would be fruitful.

3

ELECTRONIC MUSIC

In this chapter, we move back in time from the 1948 transistor period, in order to trace some of the innovations that occurred during the development of electrical and electronic music, whereby music could be generated *electrically,* using radio set loudspeakers, instead of mechanically, as, say, by organ pipes.

To make the discussions of these developments more readily understandable, let us first review some of the fundamentals of musical sounds and the underlying principles of sound production and propagation.

MUSICAL TONES

We know sound to be a pressure disturbance transmitted through the air as a wave motion. Just as the disturbance on the surface of a lake caused by a stone thrown there moves out in all directions in the form of ripples or waves, so the boom of a bass drum is transmitted in all directions in air as a train of sound waves.

A musical sound or tone differs from a random sound in that the sound waves which it generates are periodic; that is, each successive wave is similar to the preceding. The rate at which these waves repeat themselves determines the pitch of the musical note. If this rate of repetition is slow, the note has a low pitch; if the rate is fast, the note has a high pitch. When a succession of sharp pulses, or clicks, is repeated *very* slowly, the listener hears each click, and no perception of pitch occurs. As the rate of repetition is gradually increased, the sensation of *tone* is soon reached, and, as the rate is increased further, the pitch of the note is observed to rise. When the repetition rate is steady, the succession of pulses is *periodic,* and one then perceives a musical tone.

This succession of sharp pulses is the basic method of tone production of almost all orchestral instruments. In the case of the violin, as the bow is drawn across the string, the string is drawn along with it for a short distance until the sticky, rosined, bow can no longer hold it, and the string suddenly slips, to be caught anew by the bow, and the process is repeated. This occurs so rapidly that the eye cannot follow the motion, but each pulse produced by the slipping of the string is transmitted to the sounding mechanism of the violin, and the rapid succession of pulses is heard as a musical tone.

Similar pulses are produced in brass instruments by the motion of the performer's lips. As he blows into the mouthpiece, his lips are in rapid, vibratory motion. Once during each vibration, the lips come together suddenly and shut off the wind supply. The periodic series of jolts or pulses thus produced in the wind supply is heard as a musical tone. In reed instruments, such as the woodwinds and reed organ pipes, a similar shutting off of the wind supply is effected, and the rapid succession of these pulses constitutes the basic reed tone. Even the basic tone of the human voice is nothing more than a series of pulses generated by the vocal chords periodically shutting off the wind supply from the lungs. Each of these several methods of basic tone production can result in quite different tone qualities, where by tone *quality* we mean the characteristics of a tone which distinguish it from others of the same pitch and loudness. Thus a violin tone and oboe tone of the same pitch and loudness differ because they have different tone qualities.

Returning again to fundamental principles, let us recall that the actual sound consists of a disturbance of some kind produced by a moving object. In the case of a drum, it is the motion of the stretched membrane that initiates the sound waves. In the case of the violin, it is the vibrating body of the violin which causes the sound. In the case of the electrical device known as a loudspeaker, the sound is produced, as in all radio sets, by the motion of the paper cone of the device. The purpose of a loudspeaker is to convert electrical waves into sound waves. Thus, in *any* loudspeaker, whether it be in a radio set, theater sound system, or any other, the sound waves emanating from it were produced by electrical waves possessing the same characteristics of pitch (or frequency) and tone quality (that is, wave form). By magnetic means, the paper cone is made to move to and fro in exact accordance with the electrical waves impressed upon the loudspeaker terminals, and air disturbances are set up producing sound waves corresponding to those electrical waves. To produce music electrically, then, we first generate electrical waves by any of several methods and then convert these waves or oscillations into sound waves by means of the loudspeaker.

HARMONIC ANALYSIS

In 1862, the German scientist Hermann von Helmholtz showed that any periodic tone consists of a single "fundamental" tone and a plurality of overtones, whose pitch or frequency is an exact multiple of this fundamental tone. Figure 11 shows an analysis of a sound consisting solely of the fundamental sound, and Fig. 12 shows one of a sound comprising a fundamental and numerous overtones (all multiples of the fundamental). Thus a tone which we recognize as having a pitch of 200 cycles per second (Fig. 12) can often comprise, in addition to the 200-cycle tone, other component tones possessing frequencies of 400, 600, 800, 1000 cycles, and so on. Helmholtz furthermore showed that the tone quality of any note was determined by the relative strengths of these overtones or harmonic components. For example, if the note analyzed in Fig. 12 had had weak overtones it would have possessed a round, fluty quality, whereas if its overtones were strong it would have possessed a rich, violinlike quality.

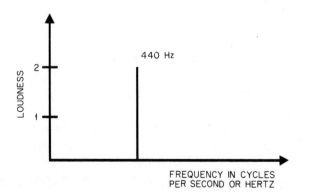

Fig. 11. Representation of a single-pitch, single-frequency tone. The position of the vertical line along the horizontal (frequency) axis shows the pitch of the tone, and the height of the line indicates the amplitude or loudness.

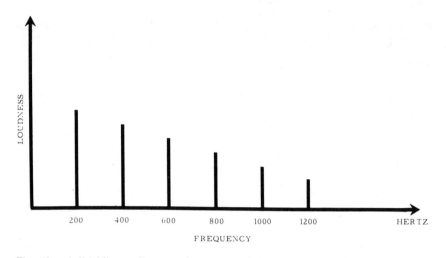

Fig. 12. A "rich" sounding tone has numerous harmonics, in addition to the single, fundamental tone of Fig. 11. They are all multiples of the fundamental.

Physics textbooks, in their section on sound, invariably include a discussion of the Helmholtz analysis procedure of Fig. 12, and some leave the reader with the impression that a single diagram of that type, indicating the position (pitch) and height (loudness) of each component, is sufficient to indicate the tone quality of a given musical instrument. It took a concept involving a property called a "formant" to show that such a single analysis of harmonics (overtones) is *not* adequate for specifying the tone quality of the instrument.

We shall discuss the subject of formants at some length shortly. Let us at this point, however, review some of the early developments in the electrical production of music.

THE FIRST ELECTRIC ORGAN

It was shortly after the beginning of the twentieth century when Thaddeus Cahill devised a way of building an electric organ. He recognized that because the rotating electrical generators by which power companies furnished many homes with 60-cycle house current could also be designed to generate other frequencies, an assemblage of such generators could act as a source of musical sounds. Each generator would produce an alternating current having a frequency corresponding to the frequency of a particular note of the musical scale. When the organ key corresponding to that particular note was depressed, it would close an electric switch, thereby connecting the rotating power generator to a device similar to a present-day radio set loudspeaker. This device would then transform the constant-frequency alternating *electric* current into a constant-frequency (constant pitch) *sound* signal, and when several keys were depressed an organlike *chord* would be sounded!* Cahill named his instrument the "Telharmonium" [Fig. 13(a)].

Because the tones from single generators would contain only the fundamental—that is, no harmonics would be sounded—Cahill arranged for *combinations* of tone generators to be available to the performer. These combination tones would have both odd and even harmonics added to the fundamental so that the tone quality of many musical instruments could be approximated. However, we shall see later that most orchestral instruments (and also the human voice) possess a tone quality in which only those harmonics in a certain pitch region are accentuated. This characteristic, called a *formant,* is practically impossible to attain with Cahill's procedure over a wide range of fundamental pitches. Cahill did, however, recognize another important problem, which he called "robbing,"† that occurred when one generator was called into use in several places, causing its contribution to each place to be undesirably lessened, that is, to be "robbed." He overcame this robbing by inserting impedances in each of the many leads coming from each generator. Many present-day electronic organs also use this procedure, first described by Cahill in his 1908 patent application.

Because the radio tube had not yet been invented, no *amplifiers* were available to Cahill, so his rotating electrical generators had to be made rather powerful (and hence rather large) in order to cause the sound issuing from his electric-to-acoustic converter to be loud enough. The organ was therefore a very bulky affair. Nevertheless, people traveled great distances just to hear this new musical instrument, the first electric organ.

Many years later, the electric clock expert Laurence Hammond used tiny rotating generators, radio tube amplifiers, and modern loudspeakers to create the first really practical electric organ, and today

---

* *Electrical World, 47,* p. 519, and *48,* p. 637 (1906).
† Robbing was described in Cahill's application No. 436,013 (filed June 1, 1908) and in his application No. 485,645 (filed March 25, 1909).

(a)

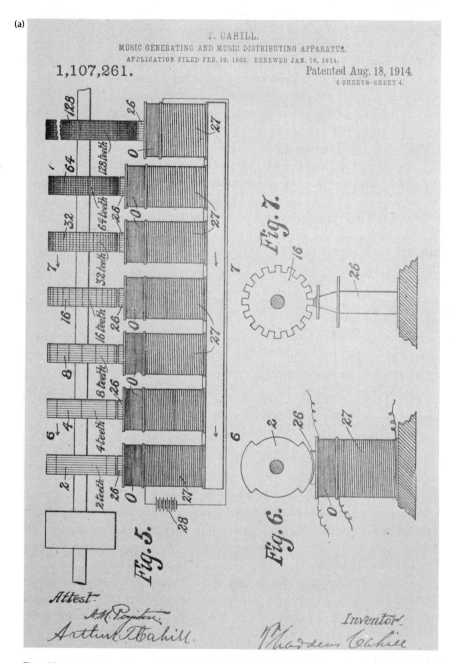

Fig. 13. (a) A page from Cahill's 1914 patent on his Telharmonium, and (b) a page from the 1934 patent on the Hammond organ. The much greater compactness of the latter is evident.

(b)

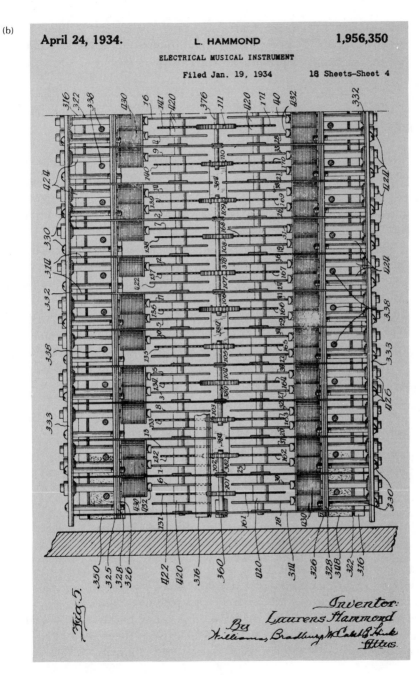

April 24, 1934.

L. HAMMOND

1,956,350

ELECTRICAL MUSICAL INSTRUMENT

Filed Jan. 19, 1934

18 Sheets—Sheet 4

Fig. 5

Inventor:
Laurens Hammond
By
Williams, Bradbury, McCabe & Hink
Attys.

models of this organ are found in many thousands of homes and churches. Figure 13(b), from the very important Hammond patent, shows the striking compactness of the electric-clock technique, also the application filing date and the date of issue.

THE EARLY DAYS OF RADIO

Cahill's instrument made its debut long before today's technology of radio and electronics was born, and even today its rotating generator signal production process would not, in general, be called "electronic." The true radio and electronics era arrived more than a decade later, during the 1920s.

The discovery and demonstration of radio waves by the German scientist Heinrich Hertz, in 1887, and the development of the radio "tube" by the U.S. scientist Lee DeForest, 19 years later, led to a very rapid growth of the technology of radio, to be followed much later by another use of radio waves, television, first black and white, then color.

During the very early days of radio, the least expensive version of a radio receiver was one called a crystal set. It permitted a listener wearing earphones to hear music, news, and other programs broadcast from radio stations many miles away, and he usually could tune in each of the several different stations. Many young students (myself included), often of grade school age, became fascinated with the idea of making their own crystal sets, as the instructions for such a task were widely published. Timothy Leland of the *Boston Globe* related one of my experiences in those early days with the following (rough) quotation:

> I remember one day my mother came into our living room and found my father, my brother and me lying on our backs on the floor, staring up at the ceiling. I guess she thought we were dead or something because she let out a little shriek. All we were doing was listening to my crystal set.*

To make the crystal sets of those days, the young entrepreneur was told how to wrap copper wire around a (cylindrical) oatmeal box, how the crystal detector and the headphones were to be connected, and so on. Successful completion of the set gave these young builders a tremendous feeling of accomplishment, and to their parents they seemed like true geniuses, because they themselves could hear from the handiwork of

* T. Leland, *The Boston Globe* (September 7, 1964).

their offspring, that little crystal set, music coming from hundreds of miles away.

Soon, however, these young enthusiasts felt that they should aspire to a greater accomplishment, that of building a radio set having a *vacuum tube* in it, in place of the simple crystal. This task was much harder, as it involved providing two kinds of battery supplies (a so-called A-battery and a B-battery), many more components, and many more complicated connections. Nevertheless, this new age of innovation-conscious pioneers usually succeeded, and then their next move was to incorporate in the set several vacuum tubes, so that a *loudspeaker* could be used (instead of the headphones), thereby permitting everyone in the room to hear music originating from distant cities.

RADIO AND ELECTRONIC MUSIC

The early Webster defines *serendipity* as the discovering of things not sought, and it was at this point in time that serendipity occurred in connection with electronic music. The early home-made vacuum tube sets, particularly those called "super-heterodyne sets," often possessed the annoying habit of suddenly emitting a loud squeal from the loud-speaker, causing the embarrassed young designer to leap immediately toward the set to readjust the tuning and thereby stop the squeal. In the process of his approaching the set, the nearness of his body introduced a (capacitative) electrical effect which caused the *pitch* of the squeal to change, going either to a higher-pitched tone or to a lower-pitched one.

Now these squeals were not very musical, so that the step from radio squeals to electronic music was not too obvious. The radio set had been constructed to receive radio programs, and the connecting of these squeals with *electronic music* involved both serendipity (discoveries not sought) and creativity. But several ingenious people did realize that a *tone* was being produced by the radio set and loudspeaker, and they accordingly began to reflect on the possibilities of generating, electroni-cally, tones which would be more musical than that radio set squeal. They thereby became the first to recognize that the field of music might benefit from the technology of radio.

One of the early electronic music tone generators employed an effect identical to that just described. Just as the pitch of the squeal in the radio set was altered by changing the proximity of a person's body

Fig. 14. One version of a very early electronic musical instrument, the Theremin. The pitch of the tone generated is determined by the proximity of the performer's hand to the vertical conducting rod.

to the set, the pitch of the tone produced by this electronic musical instrument (called the "Theremin," after its inventor, the Russian scientist Leo Ssergejewitsch Theremin) could be controlled by the proximity of the performer's *hand* to a metallic structure (usually a rod) connected to the instrument. In addition, a *wavering* motion of the performer's hand could generate a tone of wavering pitch, similar to the vibrato or tremolo of, say, a trombone tone produced when the trombone slide is made to move back and forth at a speed of five or six times a second. The pitch was low when the hand was far distant, and when the hand was close the pitch was high.* A range of pitch of several octaves was available.

This instrument received much attention in the United States, possibly because of the novel and striking method of operation. It utilized two vacuum tubes oscillating at radiofrequencies with the beat note between them giving the audio signal. A small percentage change in the frequency of the oscillators themselves produced a large percentage change in the frequency of the audio beat note, so that body capacity effects alone were sufficient to vary the beat note frequency through the full audible region. The body capacity effect was introduced

* P. Lertes, *Elektrische Musik* (Theodor Steinkopt, 1933).

by varying the proximity of the hand to a vertical conducting rod connected to the tuned circuit of one of the oscillators. A similar rod (and the capacity effect) was used to vary the *volume* of the generated tone. Figure 14 shows a version of this instrument (the box with the rod) which could be attached to a radio receiving set. The rear box, with its "1933 modern" design, houses the loudspeaker.

Other early electronic musical instruments used keyboard-actuated switches to change the pitch in steps. Often these were attached as separate keyboards to a standard piano, which could then act as an accompaniment instrument for the electronically generated sounds.

To show that even in 1933 electronic music was already well established, a translated excerpt from the preface of Lertes's *Elektrische Musik* is quoted here:

> Even today, electrical music is looked upon by the majority of performers and composers as an intrusion into a cultural and intellectual realm which apparently possesses no place for technical science, an intrusion into an artistic sphere of activity which has already experienced a significant restriction of its possibilities through the phonograph and radio. This view toward electrical music is fundamentally wrong. The phonograph and the radio are technical accomplishments that have actually made the music realm more *passive*. Electrical musical *instruments,* in contradistinction, return music to an *active* practicality, for these instruments are, and will remain, art instruments.

ELECTRONIC DEVELOPMENTS

As we know now, "electrical" (electronic) music did not stop with these elementary instruments. Soon complete chorus instruments, the electronic organs of today, were developed, permitting thousands of churches and homes to have at a fraction of the cost and size of equivalent pipe organs, many-voiced, many-tone-color, instantly responsive electronic musical instruments.

To provide a separate tone for each key, so that many-tone chords could be played, a separate electrical tone source is required for each key (just as a separate organ pipe is required for each key of a pipe organ). But in the electric organ case, one electronic tone source can serve as a producer of many *different* timbres or tone colors, because the electrical vibrations of the tone generator can be passed through

Fig. 15. The author playing an experimental model of his first organ-type electronic instrument. It employed dozens of tiny neon tubes as tone generators and could therefore provide many-toned (choral or polyphonic) music, as opposed to the instrument in Fig. 14, which could provide only single-toned (monophonic) selections.

electrical circuits which can modify the tone color of that electrical source when it is finally transformed into an acoustic tone by the loudspeaker. Thus, whereas in a pipe organ a complete rank of 70 or so pipes is needed for *each* tone color (e.g., 70 pipes for the diapason tone, 70 for the flute tone, 70 for the trumpet tone), only one set of 70 electrical generators is needed in an electronic organ, because many different qualities of tone are available from this one set of electronic generators.

However, in the very early days of electronic music, even this significant advantage still posed a problem, because the use of standard radio tubes as tone sources meant that 70-some tubes, plus all of the associated circuits, were required, and this involved an appreciable expense (even the finer radio sets in those days had only six or eight radio tubes).

Accordingly, an early experimental electronic organ that I constructed as partial fulfillment of my electrical engineering degree thesis used tiny *neon* tubes as its tone sources, since these were quite inexpensive compared to vacuum tubes. This organ, first demonstrated in 1932, is shown in Fig. 15. As the front-page location of this newspaper account of the instrument suggests, the concept was looked upon as being very newsworthy. Because I had studied piano and organ (Appendix 3 and 4), it was decided that I should demonstrate this experimental organ at the 1932 Engineering College open house day, which was celebrated each year at the University of Cincinnati.

But this instrument had advantages and disadvantages. As I reported at an Institute of Radio Engineers meeting held in Cincinnati,

> The small size and small cost of the neon tube is one advantage of this type of instrument. However, the fact that many tone colors are obtainable with one tube through the use of formants is perhaps an even greater asset. One serious difficulty is that of pitch maintenance. The frequency of oscillation depends upon several factors and changes in any of these factors will affect the pitch, so that retuning is necessary. With a single voice instrument, such as the Theremin, this is not so serious, as the performer can detect the variation and correct for it. When many neon tubes are involved, however, the bringing of 70 or so neon-tube oscillators back into proper pitch is quite another matter. In the neon tube organ, which I demonstrated on Co-op Day last year, several steps were taken to correct for this difficulty. However, even with these precautions, pitch variations were in evidence, and retuning was necessary before every performance.

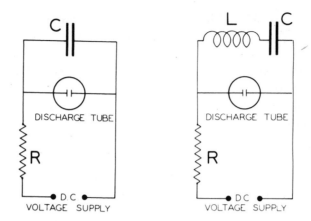

Fig. 16. The "standard" intermittent glow discharge oscillator circuit (left) and the "inductive" circuit (right).

PITCH STABILIZATION

For my master's degree thesis, I undertook a research program aimed at learning more about the neon tube oscillator properties, that is, about the intermittent glow discharge. Figure 16, taken from that thesis, shows the pitch stabilization problem, including one way in which an inductance can be added to the standard intermittent glow discharge oscillator circuit. With this procedure, it was hoped that a "resonance" effect could be introduced, with the resonant frequency being determined (as in other circuits containing an inductance and a capacitance) by the original capacitance and the newly introduced inductance. As seen in Fig. 17, this addition of an inductance did indeed result in a stabilizing of the oscillation frequency. In the region $X–X$ of the lower curve, B, a sizable variation in the supply voltage (one which, as seen on the upper "noninductance" curve, A, would cause a very large variation in pitch of the oscillator signal) caused practically no change in the oscillator pitch. These results were the subject of a paper published in the journal *Physics* in October 1933,* shortly after I had received my master of science degree. This paper was the second of my publications, the other having appeared several years earlier in the more popular journal *Radio News*.

* W. E. Kock, "The Effect of Inductance on the Intermittent Glow Discharge," *Physics, 4,* pp. 359–361 (October 1933).

DOCTORAL RESEARCH

In the spring of 1933, I received word that I had been awarded an Exchange Fellowship for continued study at the University of Berlin, with arrangements for me to conduct my doctoral research at the Heinrich Hertz Institute at the Berlin Technische Hochschule (the equivalent of our engineering colleges) under its Director, Karl Willy Wagner. In discussions relating to this continued research, Professor Wagner recalled that three particular forms of oscillations had been observed when an inductance and condenser were inserted in the circuit of a powerful *arc light* and that the finding of three similar forms of oscillations in the inductive *glow discharge* circuit would constitute an excellent subject for a doctoral thesis. Positive evidence of these three types was found, and a paper* describing this research was published in the October 1934 issue of the German journal *Zeitschrift für technische Physik* (which translated is *Journal of Technical Physics*).

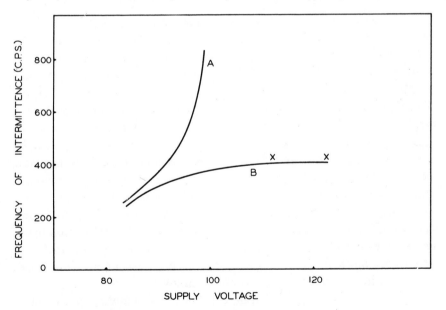

Fig. 17. Curves showing the variation of pitch with applied voltage of the standard glow discharge circuit (upper curve A) and of the inductive circuit (lower curve B). In the region X–X, the change of pitch with voltage is extremely small.

---

* W. E. Kock, "Der induktive Glimmentladungsoszillator," *Z. Tech. Phys., 19,* pp. 377–384 (October 1934).

Although I had been cautioned that the bestowing of a Ph.D. degree on a student who had spent only two semesters at the University of Berlin was exceedingly unlikely, I decided to try for it anyway. I chose for my examiners in my two major fields, theoretical and experimental physics, Professor Max von Laue (a Nobel Prize winner) and Professor Arthur Wehnelt (known for his improvements in cathode-ray tubes). For my two minor fields (mathematics and philosophy), I chose as examiners Professor L. Bieberbach and Professor Wolfgang Koehler. As Appendix 5 indicates, my examinations were successful (rated cum laude) and my thesis was accepted (rated laudable). The date of the award of my doctorate in physics was set for December 1934, but because I had earlier accepted a one-year Teaching Fellow scholarship at the University of Cincinnati, I had to accept the degree in absentia.

THE FORMANT CONCEPT

We now turn to the subject of formants, tonal properties which are often important in determining musical instrument tone quality. *Formant,* a German word meaning, roughly, "that which forms," was first employed extensively* by Professor Karl Willy Wagner during the late 1920s. (This was the prime reason for my request to conduct my Exchange Fellowship research under Professor Wagner.) Wagner's analysis showed that formants were responsible for the ability of the human voice to create its many different voiced sounds, such as the *oo* sound, the *ah* sound, and the *ee* sound. Wagner's analysis has now become fully accepted as the way for describing the vocal resonances that are responsible for the tone quality of voiced sounds. A little later, creativity entered the formant scene when scientists realized that musical instrument sounds *also* contain formants. This realization led to interesting uses of electrical resonant circuits to produce, *electrically,* in electronic musical instruments, the musical instrument resonances contained in true orchestral instrument sounds.

Figure 18 shows three different harmonic analyses of the same instrument, a bassoon, all of these being comparable to the analysis of Figure 12. The top diagram shows the harmonic structure of a high-

---

* K. W. Wagner, *Elektrotechn. Z., 45,* p. 451 (May 8, 1924); see also *Sitzung Berichte Akademische Wissenschaft, Phys.-math.,* Kl, Bd. 1, p. 1. (1933).

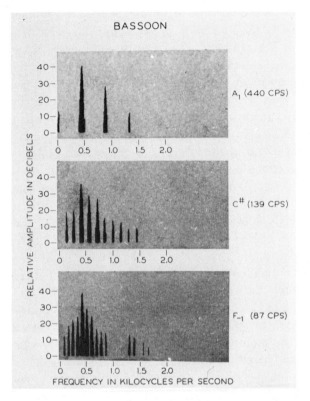

Fig. 18. Three harmonic analyses of one orchestral instrument, a bassoon, show that strong variations in the strength of the individual harmonics occur when the pitch is changed (from A to C# to F).

pitched note of the instrument. The tallest line, at the left of the diagram, corresponds to the fundamental, and the two higher harmonics are lower in intensity (loudness). In the middle diagram, the tone sounded has a lower pitch; for this note, the third vertical line (the third harmonic) is the tallest, and several others of the harmonics are also seen to be stronger than the first line, the fundamental. In the bottom diagram, the pitch of the note being played is quite low, and now there are large numbers of harmonics which are louder than the fundamental. It would obviously be difficult to say which of these three Helmholtz harmonic analyses of a bassoon best typifies the "tone quality" of the instrument, since all three (and any others of different pitch than those shown) are completely different.

However, when one calls upon the formant or resonance concept to define or specify the tone color of the trumpet, one is much better off. We see from Fig. 18 that *all* of the three analyses show the loudest components to be located approximately at a pitch of 440 cycles per second (440 hertz). A resonance or formant located at this point in frequency can thus account completely for the difference in these three analyses. We can thus say with some confidence that the tone color of the bassoon is largely dependent on a resonance, a formant, a tone-color "former," located at about 440 hertz in the musical scale.

Because of the importance of a knowledge of the characteristics of the human voice, particularly in the science of voice communications (such as by telephone), it is not surprising that the first area where a systematic analysis of tone quality by Wagner occurred was one involving *voice* tone quality analysis. As noted above, Wagner showed that voiced sounds contain strong resonances. We are familiar with the very low-pitched resonances which can be produced with the voice by shouting, for example, into a barrel. The resonances of the voice are, like the resonance of Fig. 18, in the much higher frequency region. These resonances are due to the existence of various cavities between the vocal chords and the mouth. Their small size (relative, for example, to the low-frequency barrel) accounts for the high frequency of the resonances. Different mouth positions, causing the different cavities to acquire different sizes, change the frequency (pitch) of the resonances, and Wagner showed that it is this change in the frequency of the formants which changes the sounds of our voices (as from the *ooh* sound to the *ee* sound).

VISIBLE SPEECH

The most effective procedure for showing the importance of resonances is by means of a technique called "visible speech," developed by the U.S. scientist Ralph K. Potter at the Bell Telephone Laboratories shortly after World War II.* In this analysis procedure, by plotting the frequency or harmonic analyses of Figs. 11 and 12 versus time, *changes* which occur with time in the sounds can be portrayed in the final

---

* R. K. Potter, G. A. Kopp, and H. C. Green, *Visible Speech* (New York: Van Nostrand, 1947); see also W. E. Kock, *Seeing Sound* (New York: Wiley, 1971).

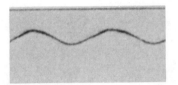

Fig. 19. The time representation, by the Potter sound spectrograph, of a single-frequency, sinuous (sinusoidal) wave. Time runs from left to right, pitch runs vertically.

record. Figure 19 shows a visible display of a single-frequency tone, such as that analyzed in Figure 11, where the pitch is varied with time. In this figure, time runs from left to right, and pitch or frequency is plotted vertically, so that the pitch is indicated by the varying *height* of the wavy line shown. Figure 20 shows a tone rich in harmonics (such as the one analyzed in Fig. 12) also being varied, up and down, in pitch.

In Fig. 20, all of the individual harmonic components are portrayed. Potter soon realized, however, that by using a very broad analysis filter having a frequency band which extended over a number of harmonics the formant positions would be much more clearly portrayed, particularly for the important analyses of voice sounds. Figure 21 shows two (unchanging) records of the vowel sound of the *a* in "at." The record on the left was spoken by a man and that on the right by a woman. The higher pitch of the woman's voice would have made the *harmonic* analysis (as in Fig. 12) of the two quite different, but the similar *formant* positions permit the analyst to recognize that the sound really is the same vowel sound in both cases.

Figure 22 portrays a number of vowel sounds, showing how the positions of the formants or *bars,* as Potter often referred to them, vary for different vowel sounds. Figure 23 shows the analysis of a complete sentence, "This is the news." The *s* sound in the word "this" shows its

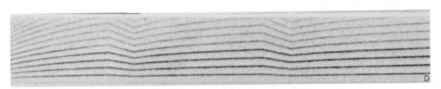

Fig. 20. A representation or spectrogram, similar to Fig. 19, except that here the tone is rich in harmonics. It rises slowly in pitch, then descends more rapidly, repeating this pattern.

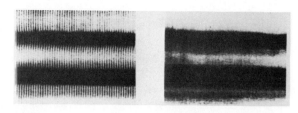

Fig. 21. When broad-band (wide frequency band) filters are employed in the analysis of sounds, only broad frequency regions are accentuated. Here two voice spectrograms of the same vowel sound are shown: the left as spoken by a man, the right by a woman.

strong high-frequency, noiselike property. But the variations in the formant positions in the right portion of the record demonstrate the value of the Wagner formant concept and the Potter visible-speech portrayal procedure.

FORMANTS AND HALLFORMANTEN

In 1931, my father (both he and his father were also patent conscious, as seen from Appendix 6 and 7), informed me of a 1930 publication* in which the author, F. Trautwein, discussed (in German) the concept of electrical (electronic) music and also set forth a new theory on tone quality. This theory involved formants, with Trautwein using the term *Hallformanten,* the German for "sound [*Hall*] formers [*Formanten*]." His theory reads as follows:

> The physiological effect of musical quality is, in the main, produced by the presence of one or more "Hallformanten" which are heard simultaneously with the groundtone [the fundamental]. These are damped oscillations of a definite frequency which are invariably higher than that of the fundamental and which may be in any relation to this, i.e., not necessarily a multiple. The Hallformant invariably dies during each period of the fundamental, or is suppressed by the beginning of the next.

The prior theory, set forth by Ohm in 1843, and later amplified by Helmholts, stated:

> All musical sounds are periodic; the human ear perceives pendular vibrations alone as simple tones. All varieties of tone-color are due to particular

* F. Trautwein, *Elektrische Musik* (Berlin: Weidmann, 1930).

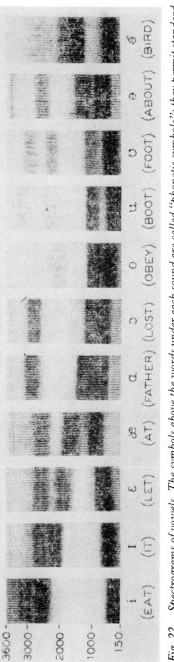

Fig. 22. Spectrograms of vowels. The symbols above the words under each sound are called "phonetic symbols"; they permit standard vowel sounds to be more simply identified.

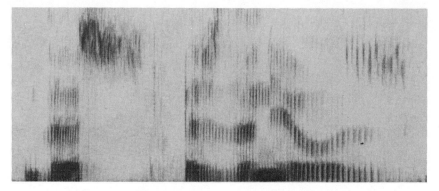

Fig. 23. A spectrogram of the words "This is the news."

combinations of a larger or smaller number of simple tones; every motion of the air which corresponds to a complex musical tone or to a composite mass of musical tones may be analyzed into a sum of simple pendular vibrations and to each simple vibration corresponds a simple tone which the ear may hear.

As to the tone quality imparted by Hallformanten, Trautwein noted:

> Small variations in the characteristics of the Hallformant produce noticeable changes in the tone quality. The frequency of the Hallformant affects the tone quality as follows: a low pitched Hallformant gives a hollow, bassoon-like tone; in the middle register, Hallformanten impart a tone quality similar to the clarinet; high Hallformanten give a sharp quality, similar to a trumpet. The decrement of the Hallformant, i.e., the degree of damping, also affects the tone quality. Weakly damped Hallformanten impart a shrill quality, whereas heavily damped ones produce a round mellow tone.

Now it turns out that several phenomena in sound are easily explained through Trautwein's theory. For example, when the speed of a record player is changed, the quality of the tones recorded is significantly modified. Voices are changed so completely that they are not recognized as those of the original speaker. This is explained by the variation in pitch of the Hallformant caused by the higher speed of the playback of the recording. (Recall that for any *fixed* tone quality the Hallformant remains fixed in pitch over the entire scale of the instrument.)

ENGINEERING THESIS

At the time of the arrival of Trautwein's publication, I was seeking a research subject for my electrical engineering thesis. After studying the publication, I decided to enter the field of electronic music, with, as an additional, related research project, an investigation into which of the two theories (Ohm's or Trautwein's) was superior. Because, in Trautwein's theory, the validity of the Fourier analysis phenomenon in the ear was questioned, it implied that the ear grasps the Hallformant itself as a single determining effect on the senses rather than analyzing it into its components. Attention was therefore directed to the determination of whether the ear does or does not analyze the Hallformant into its components.

The method employed was as follows: Produce a sound wave of audible frequency and superimpose on it a Hallformant of such a frequency that the two waves are harmonically unrelated. For example, choose a fundamental frequency of, say, 300 cycles and a Hallformant frequency of 1000 cycles. A Fourier analysis of the (constant-pitch) periodic wave will show components of 300, 600, 900, and 1200 hertz, and so on, but no 1000-hertz term. Then if a Helmholtz resonator tuned to 1000 hertz gives a noticeable effect when the above sound wave is tested the conclusion can only be that the ear does not analyze the tone into its components but grasps the 1000-cycle Hallformant as a single determining factor in the physiological effect on the ear.

The circuit of Fig. 24 was devised for this investigation. A fundamental of approximately 68 hertz was produced by the neon tube and an oscillator at the left caused an (electronic) Hallformant of approximately 370 hertz to be excited in the circuit on the right. The frequency of the Hallformant was obtained by first causing the neon tube circuit to generate pulses at a frequency of 1 or 2 cycles per second. The signal from the loudspeaker then sounded like a repeated, dying-out piano tone, so that the Hallformant frequency could be determined by the use of either a Helmholtz resonator or a pitch pipe. Figure 25 portrays such a "dying-out," repeated Hallformant signal.

The oscillator was then adjusted to give the proper frequency (68 hertz) for the groundtone, and the resulting output tone was analyzed by means of (variable) Helmholtz resonators. As the resonators were varied, the volume of sound issuing from the resonator earpiece

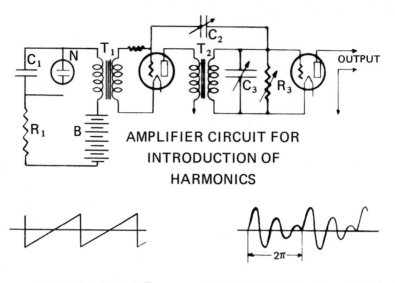

AMPLIFIER CIRCUIT FOR
INTRODUCTION OF
HARMONICS

NEON TUBE WAVE   SUPERIMPOSED TRANSIENT

Fig. 24. The circuit used to compare the formant (Hallformant) theory and the Ohm–Helmholtz theory. The circuit on the right, containing an inductance and a condenser, electronically generated the formant.

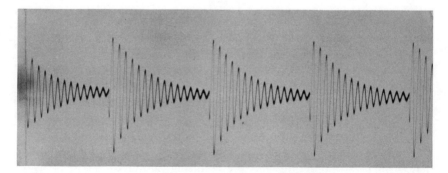

Fig. 25. A periodic series of pulses, as generated by the left-hand portion of the circuit in Fig. 24, produces repeated formants, sine waves which decay in amplitude with time.

increased whenever its position in frequency corresponded to a harmonic of the groundtone. Increased resonator output was found to occur

at 272 hertz (the 4th harmonic)
at 340 hertz (the 5th harmonic)
at 408 hertz (the 6th harmonic)
at 476 hertz (the 7th harmonic)

and so on. Special care was taken in attempting to perceive an effect in the resonators in the neighborhood of the Hallformant frequency (370 hertz), but none could be detected.

From this evidence, the conclusion was drawn that the ear does not analyze the wave into its groundtone and a Hallformant but rather hears each single harmonic and *only* the harmonics of the fundamental. In other words, the ear does not accept the Hallformant as a single determining factor upon the senses, but, on the contrary, it analyzes it according to Ohm's theory into frequencies which are harmonically related to the fundamental frequency.

As a second verification of the validity of the Fourier analysis concept, a "discrete" Fourier analysis *table* was designed (Fig. 26). Such tables, first described by Runge,* greatly reduce the number of calculations when (discrete) *samples* of a repeated waveform (Fig. 27) are used to determine the amplitudes of the harmonic components of the periodic wave. Thus Lipka† has noted that in determining the amplitude of six harmonics by the use of such a table the time and labor spent are less than required for determining one harmonic by the direct process. Thus, Runge, in 1903, provided the first speedup of the Fourier process, a *faster* Fourier analysis (Fourier transform). Programming methods for today's high-speed computers, patterned somewhat along the lines of Runge's tables, were devised by Cooley and Tukey‡; their procedure is now universally referred to as the "fast Fourier transform" or, briefly, the "FFT."

When the newly designed table was applied to give the 370-hertz Hallformant as generated in the circuit of Fig 24, it was found that the , strength of the harmonics did indeed follow the intensity pattern of a

* G. Runge, *Z. Math Phys., 17,* p. 443 (1903).
† J. Lipka, *Graphical and Mechanical Computation* (New York: Wiley, 1921) p. 184.
‡ J. W. Cooley and J. W. Tukey, *Math. Comp., 19,* pp. 297–310 (1966).

Schedule for Harmonic Analysis

(1) $f(x) = b_o + a_1 \sin x + a_2 \sin 2x + a_3 \sin 3x + \cdots + b_1 \cos x + b_2 \cos 2x + \cdots$

(2) $f(x) = b_o + c_1 \sin(x + \alpha_1) + c_2 \sin(2x + \alpha_2) + c_3 \sin(3x + \alpha_3) + \cdots$

Checks: $Y_o = b_o + \Sigma b_m$, $\frac{1}{2}(Y_1 - Y_{23}) = .259(a_1 - a_{11}) + \frac{1}{2}(a_2 + a_{10}) + .707(a_3 + a_9) + .866(a_4 + a_8)$

$+ .966(a_5 + a_7) + a_6$

	Y_o	Y_1	Y_2	Y_3	Y_4	Y_5	Y_6	$\mu_1 = T_6 + T_{10} =$
	Y_{12}	Y_{11}	Y_{10}	Y_9	Y_8	Y_7	Y_{18}	$\mu_2 = T_7 + T_9 =$
Sum	S_o	S_1	S_2	S_3	S_4	S_5	S_{11}	$\mu_3 = T_2 - T_4 =$
Diff.	d_o	d_1	d_2	d_3	d_4	d_5	d_{11}	$\mu_4 = T_1 - T_5 =$
		Y_{13}	Y_{14}	Y_{15}	Y_{16}	Y_{17}		$\mu_5 = S_o - S_{11} =$
		Y_{23}	Y_{22}	Y_{21}	Y_{20}	Y_{19}		$V_1 = R_1 + R_3 - R_5 =$
Sum		S_6	S_7	S_8	S_9	S_{10}		$V_2 = R_2 - d_{11} =$
Diff.		d_6	d_7	d_8	d_9	d_{10}		$V_3 = R_6 - R_8 - R_{10} =$
		S_1	S_2	S_3	S_4	S_5		$V_4 = d_o - R_9 =$
		S_6	S_7	S_8	S_9	S_{10}		$W_1 = T_6 - T_{10} =$
Sum		T_1	T_2	T_3	T_4	T_5		$W_2 = T_7 - T_9 =$
Diff.		R_1	R_2	R_3	R_4	R_5		$W_3 = T_2 + T_4 =$
		d_1	d_2	d_3	d_4	d_5		$W_4 = T_1 + T_5 =$
		d_6	d_7	d_8	d_9	d_{10}		$W_5 = S_o + S_{11} =$
Sum		T_6	T_7	T_8	T_9	T_{10}		
Diff.		R_6	R_7	R_8	R_9	R_{10}		

Fig. 26. A discrete harmonic analysis schedule.

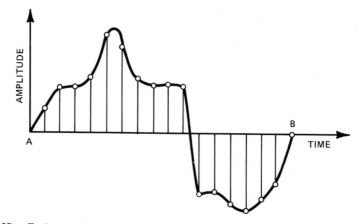

Fig. 27. Each periodically repeated waveform is sampled at discrete intervals, and when the values are inserted in the schedule of Fig. 26 a discrete Fourier analysis is obtained.

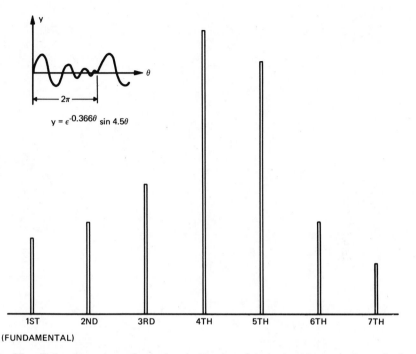

$y = \epsilon^{-0.366\theta} \sin 4.5\theta$

1ST 2ND 3RD 4TH 5TH 6TH 7TH
(FUNDAMENTAL)

Fig. 28. Taken from the author's electrical engineering thesis, this graph shows, in the upper left, a damped wave (Hallformant) similar to that in Fig. 25.

damped resonant circuit. Figure 28, taken from my electrical engineering thesis, shows, in the upper left, a damped wave (Hallformant) similar to the one in Fig. 25. The analysis shows the tendency for the strongest harmonics to cluster around the formant frequency.

Having thus completed my "related" research project, I turned my attention to the construction of an electronic organ, employing, as we discussed earlier, a separate neon tube oscillating circuit (like the one on the left of Fig. 24) for each note of the organ keyboard. I thereby constructed an organ (Fig. 15) with Hallformanten (formants) determining some of the various tone colors which could be selected. Later, in addition to my earliest patent (Fig. 29), No. 2,046,463, filed in 1933 when I was a student in Germany, I filed in 1935 a patent application describing this use of formant circuits in an electronic organ (Appendix Item 8). Several of the claims of that patent issued include the words "formant circuit."

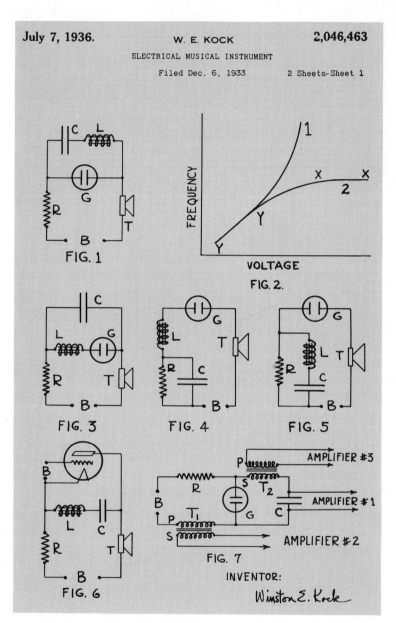

Fig. 29. This page from the author's first patent includes the drawings shown in Figs. 16 and 17.

THE BALDWIN ELECTRONIC ORGAN

With my electrical engineering thesis background, and the recognition, during my stay in Berlin, of the importance attached there to electronic music, I continued my efforts toward a commercially acceptable electric organ. The Baldwin Piano Company showed strong interest in this concept, and on the basis of the two patent applications (Fig. 29 and Appendix Item 8) a contract was arranged for royalty payments should the organ reach the commercial stage.

Some years later, commercial models of the organ did begin to appear in which the neon tube sources had been replaced with radio tubes called "double triodes," single tubes that were equivalent to *two* radio tubes. This procedure reduced the number of tubes to the fairly reasonable number of 37, thereby permitting a rather compact-design tone source having much greater pitch stability than was possible with neon tubes. Figure 30 shows the 37-tube generator of a commercial model of this organ as produced by the Baldwin Piano Company shortly after the end of World War II. Figure 31 shows the filter circuits with which the various tone colors were produced, and Fig. 32 shows the complete, assembled organ. The following excerpt is from a 1947 newspaper article by *Cincinnati Times-Star* reporter Eleanor Bell, which referred to this organ:

AN IDEA AND HOW IT GREW

A knowledge of physics and a knowledge of music proved to be a happy combination for the Baldwin Piano Co., when Winston E. Kock, possessor of said knowledge, walked into their plant about 13 years ago, his head bristling with ideas.

The Baldwin experts took Winston into the research department without much fuss and allowed him to fiddle around with his neon tubes and his loudspeakers to his heart's content. They let him knock off for a year to study with another physicist and part-time musician, name of Albert Einstein.

I happened to be especially interested in all these developments because Winston had already demonstrated his invention to me one afternoon in the physics laboratory at the University of Cincinnati. At that time he had to sit on a metal plate about six inches square, attach metal strips like handcuffs to his wrists and plug himself in on a handy wall socket.

I looked around hastily for a broom handle to pry him loose should he suddenly stiffen and turn purple, but nothing untoward happened and he began to play on a small two-octave keyboard he had rigged up. He was

Fig. 30.   The electronic generator of the first commercial model of the electronic organ
in Fig. 15. It employs double radio tubes (double triodes), so that 37 tubes provide 73
notes of different pitch, plus one vibrato (tremolo) generator.

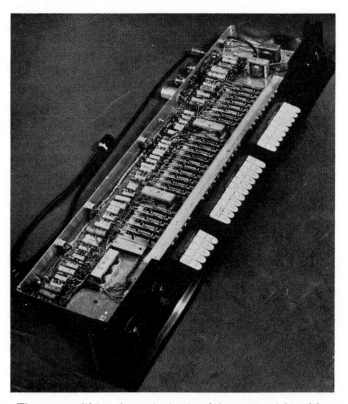

Fig. 31.   The tone-modifying electronic circuits of the commercial model contain formant circuits which are excited by the basic generators of Fig. 30.

Fig. 32. The assembled electronic organ, employing the tone generators of Fig. 30 and the tone modifiers of Fig. 31.

able to reproduce almost exactly the tone of the clarinet and oboe and cello, as I recall it.

Today the Baldwin organ has been ceremoniously presented for inspection to the Southern Ohio Chapter, American Guild of Organists, and accepted by most of the members of that erudite group as the best electronic organ to date.

It is a compact unit, no bigger than a spinet piano. The amplifiers may be tucked away out of sight. It has two manuals, standard pedals, and all kinds of recommendations that should endear it to the small and medium-sized churches for which it is intended. It won't cipher, which means that it will never break into that shrill peanut stand whistle that afflicts pipe organs.

Temperature does not affect it. Its standard radio tubes are easily replaced. It weighs 320 pounds and it costs $2650 installed, a piece of news that should cause the hearts of church music committees to quicken with joy. It uses less current than an electric iron.

There is one in use in the Mt. Auburn Presbyterian chapel, and some nice young man will play for you any time in Baldwin's downtown chambers . . . anything from a Bach Toccata to "Tea for Two."

CHURCH ORGANISTS' VIEWS

In 1947, the "organists' journal," *The American Organist,* carried a similarly favorable account, with such statements appearing in it as

> Readers may wonder how the tones of these 23 stops compare with the 23 voices of the same name in a pipe organ, and here the real value of the Baldwin electronic instrument is shown. Baldwin's engineers made analyses of these organ tones and have succeeded admirably in imitating them more closely than has been achieved to my knowledge in any other such instrument. Individually the tones are pleasing, and there is a satisfactory ensemble either with or without reeds. In short, the Baldwin is a satisfactory instrument that will please organists despite its lack of pipes.
>
> My prejudice against electrotones has been rather decided in the past. It did not seem possible that the sounds and playing-details of an organ could be approached by such a method. This Cincinnati experience changed my opinion completely. I advise all organists, whether in the market for an electrotone or not, to find out by personal experience how accurate are my words.*

The favorable reception given to the organ was largely attributed (by those who were aware of the existence of formants in musical instruments) to the fact that the organ was designed on the formant principle. This presumption was strengthened quite recently by the American Patent Law Association. As a way of commemorating the 1976 U.S. Bicentennial, this association developed a slide program involving outstanding examples of inventions and patents in order to show the workings of the patent system in the United States from a historical standpoint. Learning of this through the Cincinnati Patent Law Association, the present Research Director of Baldwin, Dan W. Martin, decided to submit the Baldwin formant organ concept as such an example, including (1) a copy of that patent showing best the formant features (Fig. 33) and (2) a photo of me playing the Dedicatory Recital on a Baldwin organ chosen by the Univeristy of Cincinnati for its Wilson Auditorium. Martin's submission was selected for the Bicentennial Program (Appendix 9), including the photograph and the patent (Fig. 33), and it was noted that "upon expiration of the patent several other producers of electronic organs adopted the formant prin-

* R. W. Dunham, "A successful electrotone," *The American Organist,* p. 158 (May 1947).

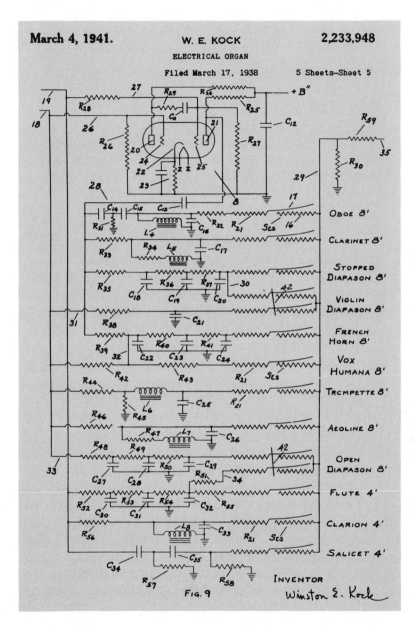

March 4, 1941.   W. E. KOCK   2,233,948

ELECTRICAL ORGAN

Filed March 17, 1938   5 Sheets—Sheet 5

FIG. 9

INVENTOR
Winston E. Kock

Fig. 33. This patent received recognition by the U.S. Patent Association in the U.S. Bicentennial Year 1976. Formant circuits are seen in the oboe, clarinet, trompette, aeoline, and clarion circuits.

Fig. 34. A modern (1975) version of the Baldwin electronic organ, using formant circuits for tone modifiers.

ciple, a tribute to Dr. Kock and to the proper operation of the patent system."

In the development of the organ, there were many who made very significant contributions, and the most outstanding of these was my classmate and colleague at Baldwin, J. F. Jordan. He is named as co-inventor of three of the early 18 organ patents and became a Vice-President and later a Director of the Baldwin Company. Figure 34 shows a recent three-manual church model of the Baldwin organ. It includes numerous features added after I had joined the Bell Telephone Laboratories in 1942, and these are all attributable to Jordan and his associates.

FORMANTS AND PIPE ORGAN MIXTURES

Before we leave the subject of formants, it is interesting to note how certain pipe organ stops make available to pipe organs various orchestral tone qualities, by simulating the presence of formants through the use of mixture stops, as I have discussed in my earlier book, *Seeing Sound* (Wiley, 1971).

Because formants have so much to do with the characteristic timbre of many musical instruments, a comparable effect is often sought in the design of pipe organ stops referred to as "mixtures." In such stops, numerous pipes are made to sound even when only a single key is depressed. For the more usual stop, such as a diapason, only one pipe speaks when one key is depressed, and hence only one set, or rank, of pipes (one for each key of the keyboard) is required. For a mixture, a plurality of ranks is needed, and the term applied to the mixture is dependent on the number of pipes that speak (for example, a "four-rank" mixture).

Mixtures that are designed to simulate formants accomplish the effect by the expedient of breaks; that is, the lower octaves have the very high partials of the depressed key represented, whereas for the higher octaves lower partials (harmonics) are made to sound. Thus, as one goes up the scale, the breaks remove upper harmonics and add lower ones. These breaks thus impart brilliance to the lower octaves, richness to the middle octaves, and fullness to the upper octaves. Helmholtz compared the resonances, or formants, of a violin to the effect of mixtures (compound stops) in the organ:

> The partial tones of the strings [of a violin] are reinforced in proportion to their proximity to the tones of the resonance box. The deepest notes of a violin will have their octaves and fifths favored by resonance, whereas the higher notes will have their prime tones assisted. A similar effect is attained in the compound stops of the organ by making the series of upper partial tones, which are represented by distinct pipes, less extensive for the higher than for the lower notes of the stop.

To illustrate how this effect is achieved let us examine a specific mixture specification. Figure 35 shows the arrangement of a three-rank cymbal mixture with six breaks; it is a stop in the Aeolian Skinner (G. Donald Harrison) organ at Christ Church, Houston. In this figure, the long continuous line represents the key that is depressed, and that note

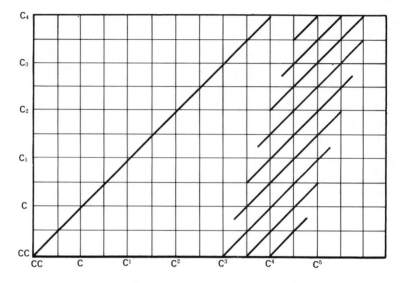

Fig. 35.   Design of the three-rank cymbal III mixture of the organ at Christ Church in Houston.

corresponds to the fundamental. Thus at the very lowest key of the keyboard, corresponding to low C (that is, CC), this three-rank mixture sounds three notes (three harmonics) encompassing C^3 to C^4. At the very highest key of the keyboard, C^4, the mixture sounds C^5, the first overtone (the second harmonic) of C^4, as well as the third and fourth harmonics. The harmonics generated lie in approximately the same region of pitch no matter where the fundamental lies, that is, between C^3 and C^6. They thus correspond to the action of a fixed resonator producing a formant. It may be noted in the chart that the harmonics shift slightly higher in pitch with a higher fundamental; this effect is also observed in the formants of orchestral instruments.

SEVERAL ELECTRICAL MUSICAL ANALOGUES

All electrical engineers must learn about electrical transmission lines, those huge towers carrying a half dozen or more thick wires seen so often all over the country. Transmission line theory involves such complex expressions as "hyperbolic cosine" and "characteristic impedance," among a multitude of others. Similarly, the analysis of the piano string has required students of musical acoustics to study very complicated theories concerning the action of the piano *hammer* on the vibrations of the piano string.

It occurred to me that, by making certain changes in the correspondences usually considered as proper in electroacoustical analysis, it might be possible to regard the piano string as being equivalent to an electrical transmission line. Sure enough, by making those changes, my electrical engineering background enabled me to devise an electrical transmission line analysis which fit the piano string theory extremely well. This development took place during the summer of a year of post-doctoral study provided by Baldwin when I worked with Nobel Laureate Sir C. V. Raman at the Indian Institute of Science at Bangalore, India. The results pleased Raman (who assisted greatly but refused the request that he be listed as co-author), and the results were useful to Baldwin in their piano research. My paper* is to appear in a forthcoming volume, *Musical Acoustics,* one of the series entitled "Benchmark Papers in Acoustics" (edited by Dr. R. Bruce Lindsay).

Another field in which electronic engineers are fairly knowledgeable is that of noise generators. Noise is always present in electrical communications circuits, and the intensity of transmitted signals must not be permitted to fall so appreciably that the ever-present noise will blot out the signal. It is like trying to hold a conversation outdoors when the noise of a jet aircraft overhead drowns out the speech signals.

To assist in the research on better communication circuits (and for many other reasons), special noise generators were developed and have been available for quite some time. Also, even though noise from these generators has a very wide bandwidth, electrical filters are known to be able to narrow down the bandwidth of the final output noise signal, so that a hiss generator (noise generator) can be made to have its noise output extend over, say, only an octave, instead of over, say, the entire audible range.

Now everyone knows that a vocal quartet or a string quartet sounds quite different than a full chorus of voices or a full string symphony orchestra. Because this "chorus effect" would be a nice tone quality to be able to reproduce electronically, some thought was given as to *why* the chorus sounded so different from a quartet singing the same

* W. E. Kock, "The vibrating string considered as an electrical transmission line," *J. Acoust. Soc. Am., 8,* pp. 227–233 (April 1937).

four-part songs. It appeared that it might be because all of the sopranos, say, were not singing exactly the same tone, so that a single frequency, as sung by one soprano of a quartet, became broadened, this effect being caused by the slight deviations in pitch as generated by the many individual voices.

To test this, the characteristics of the chorus quality were displayed by the visible-speech-portrayal technique, it being hoped to determine thereby a way of duplicating this effect in electronic instruments. Figure 36 is a record of a chorus singing the words "Thou knowest it telling" from the Christmas carol "Good King Wenecelas," made by the narrow-band analysis procedure as used in Fig. 20. The individual harmonics in the vowel sounds sung by the chorus are not nearly as sharp or so clear as they would be in solo-voice records. The existence of many voices all very slightly different in pitch produced a spreading out of the harmonics.

Because the broadening just noted is reminiscent of the broadened-frequency pattern of a noiselike sound, it suggested that an electronic-noise generator (instead of pure-tone oscillators) could be advantageous in electronic organs. This concept was tried out and proved to be successful as a way of generating a "choruslike" electronic tone. Figure 37 shows (in the top diagram) how simple electronic filters, one for each key, replace the individual electronic oscillators of the organ. They are narrow enough in bandwidth to provide a tonal quality comparable to that of the spread-out frequency bands shown in the chorus harmonics of Fig. 36. This version generates pure, fundamental tones because the filters pass only a narrow band of noise surrounding the desired note;

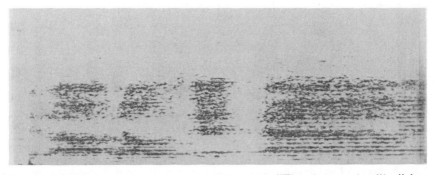

Fig. 36.   Spectrogram of a chorus singing the words "Thou knowest it telling" from the Christmas carol "Good King Wenceslas."

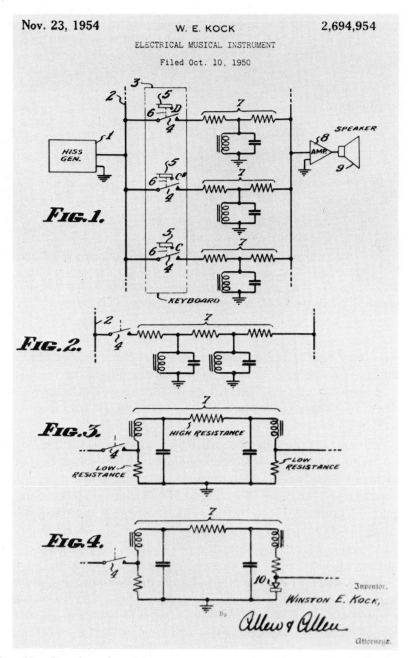

Nov. 23, 1954   W. E. KOCK   2,694,954

ELECTRICAL MUSICAL INSTRUMENT

Filed Oct. 10, 1950

Fig. 37. Procedure for employing a noise source to provide a chorus effect electronically. From a patent issued to the author.

Fig. 38. Two small radar horns on the front of an automobile provide a means for achieving a safe and automatic speed control (Bendix Corporation). This photo is from the author's book Radar, Sonar, and Holography *(Academic Press, 1973).*

the resulting tone is free of harmonics. To achieve a chorus effect with tones that are rich in harmonics, the pure, filtered tones are rectified, as shown in the bottom diagram. A large harmonic content is imparted to the tone, and these harmonics resemble those of Fig. 36 in their noise-like quality.

Still another phenomenon well known to engineers is the Doppler effect, with Doppler radars being employed in high-speed aircraft for navigation purposes, in ground-based radars for detecting oncoming enemy planes, and even (Fig. 38) in recent automobile experiments. We have all probably observed the Doppler effect for sound waves, such as when driving past a rapidly ringing bell, with the observed pitch of the bell seeming to be higher as the bell is approached and lower after passing the bell.

It was while working in Berlin for my doctorate that I conceived of a way of imparting to a constant-pitch tone (generated, say, by one of the electrical oscillators in my electronic organ) a highly desirable form of tremolo, namely a *pitch* tremolo, usually referred to as a "vibrato." A violinist generates such a pitch vibrato through the back-and-forth motion of the finger which is pressing on the string being sounded by the bow. The original specification of my first patent application (Fig. 29) included the following paragraph:

> Another method of securing a pitch tremolo effect is to have the vibrating portion of the loudspeaker move back and forth with a frequency of a few cycles per second. A listener hearing a note issuing from such a loudspeaker would observed the Doppler effect inasmuch as the sound source is moving relative to his ear. The pitch would seem higher when the loudspeaker moves towards him and lower when it moves away from him. It would thus seem that the pitch of the note would be varying and a tremolo effect would be heard. The back and forth motion of the loudspeaker could be accomplished mechanically or electrically.

Later this effect was used commercially with the loudspeaker being rotated about a vertical axis, again causing the generated tone to approach and recede from the listener.

ELECTRONIC DOOR CHIMES

During the 1930s, doorbell habits were altered by the introduction of the door chime. I was a student at the University of Cincinnati at the time (Cincinnati was where the door chime originated) and was immensely impressed, both during the early stages and later as sales mounted, with the way the Cincinnati-based NuTone Corporation handled the marketing of its new product.

All during the development of the electronic organ, it was felt that an *electronic* version of a doorbell could similarly interest the public. However, because the first electronic organs used vacuum tubes that required a certain warmup time, electronic systems for doorbells would at that time have carried the highly undesirable requirement that the device (generators, amplifier, and power supply) remain turned on at all times.

Through my later association at Bell Laboratories with the development of the transistor, I soon recognized that a transistor version would be free of that requirement. For, just as the familiar transistor radio responds the instant the turn-on switch is closed, so such a transistorized doorbell could be made to sound the instant the doorbell button is pressed by the caller.

Because this was a device that did not interest the company I joined (the Bendix Corporation) upon leaving the Bell Laboratories, a waiver was given permitting me to file for patents on various versions of the concept. At one point I ran across an article in the *New York Times* describing an installation in an elaborate home that had as a headline "The Doorbell Plays 'Aida' in a Push Button Palace." The article described how the bell at the back door played "Frere Jacques" and how a pushbutton in the bar announced martini time by playing a few measures of the "Drinking Song" from *The Student Prince*.

Because the simple substitution of transistors for vacuum tubes was not a patentable feature, I was fortunate in having recognized that one significant difference existed between the vacuum tube and the transistor versions (this involved the number of switches). Accordingly, a patent (Appendix 10) was issued on the organ variety (where the caller, on pushing the button, hears an organ chord, or a cadence, slowly dying out). This version uses, say, three inexpensive transistor oscillators (available commercially for many years) and a condenser, charged initially by the closing of the doorbell button, with its gradual discharge causing the amplifier output to die out gradually.*

A second version was also patented (Fig. 39) involving a (replaceable) tape loop of 3 to 5 seconds length, which sounds for one revolution when the button is pressed. Both would operate from standard doorbell (door chime) transformers, with power being used only when the device is activated. In his review of this patent, Marvin Camras, the well-known pioneer in tape recording, said:

> When a visitor presses your doorbell button, he is announced by a fanfare of trumpets, a Beethoven cadenza, a Christmas carol, or even your own voice. A simplified tape recorder carries these sequences on tapes wound around a cylinder. A low-voltage doorbell transformer supplies power to a transitorized amplifier and to the tape-drive motor for a complete revolution of the drum whenever the button is pressed.†

Other reports included statements like "technically known as a 'transitorized door annunciator,' to the unscientific it means a doorbell that makes music instead of noise" and "chances are the postman won't ring twice, the first ring will hold him spellbound." An amusing outgrowth of the electronic organ.

* W. E. Kock, "Electronic music in every home," *J. Audio Engr. Soc., 18*, p. 671 (December 1970).
† Marvin Camras, "Review of acoustical patents," *J. Audio Engr. Soc., 17*, p. 730 (December 1969).

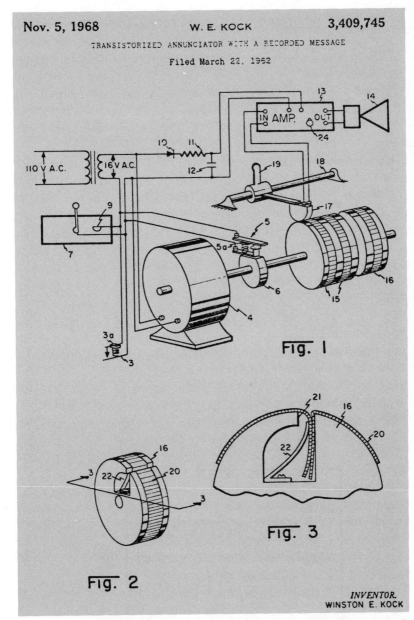

Nov. 5, 1968   W. E. KOCK   3,409,745

TRANSISTORIZED ANNUNCIATOR WITH A RECORDED MESSAGE

Filed March 22, 1962

IN AMP. OUT

110 V A.C.   16 V A.C.

Fig. 1

Fig. 3

Fig. 2

INVENTOR.
WINSTON E. KOCK

Fig. 39. Small tape loops are played back each time the doorbell button (item 3a) is pressed. The button starts the motor drive (4), and the switch (5) keeps the 16-volt doorbell power connected until the playback of the tape loop is completed. The instant response of transistors made this device possible.

RADAR AND HIGH-FREQUENCY LOUDSPEAKERS

Designers of high-quality sound reproducers, as used, for example, in high-fidelity audio systems, strive to achieve a radiated sound pattern that provides the best sound distribution for an average room, one which permits listeners located at various angles relative to the loudspeaker (right, center, or left) to experience the same sound pattern.

Because it is difficult to design one loudspeaker to handle, equally efficiently, all audible tones from the lowest to the highest, the better audio systems employ at least two loudspeakers, one for the low-frequency sounds and one or more for the high-frequency ones. The latter units are often referred to as "tweeter" loudspeakers. Now it is the high frequencies that require special attention as regards the radiation pattern just discussed. The reason for this is shown in Fig. 40. At the high frequencies, a loudspeaker becomes more *directional*; that is, it generates a narrower beam pattern at high frequencies than it does at the low ones. Accordingly, special attention must be paid to the design of the tweeter horn if it is to radiate a wide pattern for high-frequency sound, with the wide dimension of the beam pattern positioned horizontally so that a wide angular coverage (within the room) can be realized. A large *vertical* beam spread is not essential, since the listeners are usually positioned at approximately the same vertical height.

Now the pattern of a beam of sunlight, formed when the light

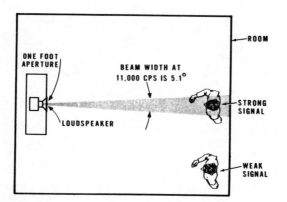

Fig. 40. High-pitched sounds issuing from a loudspeaker are concentrated into a directional beam of sound.

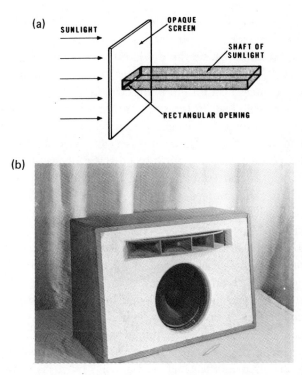

Fig. 41.   (a) Sunlight passing through a rectangular opening forms a rectangular beam. (b) The upper unit of this loudspeaker, designed for reproducing the high frequencies, has its long dimension positioned horizontally.

passes through a rectangular opening, as sketched in Fig. 41(a), retains the shape given it by the opening; in other words, the wide dimension of the beam remains lined up with the wide dimension of the opening.

Using the light beam pattern of Fig. 41(a) as a guide, designers of early loudspeakers positioned the tweeters with their wider opening or aperture horizontal, intending thereby to generate a sound pattern which was wide in the horizontal direction and less wide in the vertical direction. Figure 41(b) is a photo of such an early combination loudspeaker employing the sunlight beam orientation of Fig. 41(a) for the high-frequency loudspeaker. In this figure, the cabinet houses the low-frequency speaker (the circular opening below) and the unit above it is the tweeter.

Then suddenly radar was invented. One very early and very useful radar application was designed to assist aircraft in landing during periods of low visibility or low "ceiling." (It could be looked upon as a forerunner of our modern-day airport radars.) Weather conditions often bring about a situation where the air is quite clear (no fog) at ground level but heavy "fog" (a thick cloud layer) exists a few hundred feet above ground. In such a condition, planes can land safely but only when the pilot has available to him information regarding his position (relative to the airport runway) and information relative to his altitude, that is, his height above ground. Because this radar was able to determine accurately the plane's position and altitude, the operator of the radar could provide the aircraft pilot with the information required to land safely. This early radar procedure, involving the radar operator *telling* the aircraft pilot what he should do to land properly, was called a "talk-down" system.

For our story here, the shape and positioning of the radar's microwave radiators are what is of interest. Two identical horns having rectangular apertures were employed in this radar. The two radio (radar) beams generated by the horns were wide in the horizontal plane and narrow in the vertical plane. The two vertical beams provided information on the plane's altitude, because one of the vertical beams was aimed slightly higher than the other. The plane's distance or range from the radar was determined in the usual radar way, by measuring the time taken by the radar pulse to reach the plane and return. The two sharp vertical beams provided the radar operator with the information he needed to instruct the pilot regarding his path of descent toward the runway. If the plane was too high during his approach, his radar echo was stronger in the upper beam; if he was too low, the lower-beam radar signal was the stronger. When the pilot followed a course midway between the two beams, his position, when he finally broke through the low clouds, would be correct for negotiating a safe landing.

When this radar was first successfully demonstrated, it naturally received much publicity, and photographs of the two rectangular horns were published showing one horn aimed slightly higher than the other (Fig. 42). When most people saw these photos, they could hardly believe their eyes. Following the logic of Figure 41, they said to themselves, "Anyone knows the sharper beam of a horn is aligned with its narrow dimension, so, if sharp vertical beams are desired, why are the *long* dimensions of the two rectangular apertures oriented vertically?" Scientists who were familiar with optical matters had no such problem in understanding why the horns were oriented as they were.

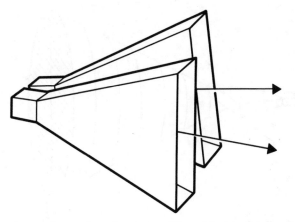

Fig. 42. Two horns of an early radar, designed to assist aircraft in landing, have their long dimensions vertical.

They knew that for astronomical telescopes the bigger the telescope the sharper the beam and therefore the more easily distant stars can be "resolved." They knew that it was the larger size of the Mt. Palomar 200-inch aperture telescope which permitted it to distinguish more clearly between closely spaced distant stellar light sources than previous smaller-aperture ones could.

This refutation of our inherent concepts on beams acquired from our experience with sunlight effects as in Fig. 41(a) is sketched in Figure 43. It shows that at some point out in front of a radiator the width of a beam is inversely proportional to the lateral dimension of the radiator; in other words, the larger the aperture, the sharper the beam. That region for which the *reversed* beam condition, that of Fig. 41(a), exists is called the "near-field" region. For sunlight, this near field can extend to great distances because the light waves involved here have extremely short wavelengths. On the other hand, for almost all cases involving audible sound waves or radar waves, this near-field region extends only an insignificant distance in front of the radiator, and therefore the relationship as sketched in Fig. 43 is the valid one.

This indicates that, in order for a high-frequency loudspeaker, a tweeter, to spread its sound in the horizontal direction and concentrate it in the vertical area (so as to provide all listeners in a room with the most desirable sound reproduction), it should have its opening (its aperture) positioned so that the longer dimension is vertical and the narrow

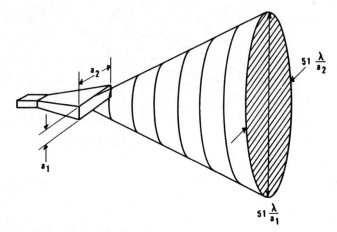

$51 \frac{\lambda}{a_2}$

$51 \frac{\lambda}{a_1}$

Fig. 43. For microwaves and sound waves, the wide aperture of the radiator generates the sharper beam.

Fig. 44. In order to spread sound horizontally in auditoriums and also to achieve directional gain in the vertical (audience) direction, loudspeakers are now arranged along a vertical direction.

dimension horizontal (just the reverse from that in which early tweeters were positioned).

Designers of loudspeakers soon realized this, and now speakers for high-fidelity systems can be purchased with tweeters having the long dimension of their apertures vertical. Rather recently, the long vertical dimension concept has been extended to *arrays* of loudspeakers, whereby a large number of speakers are placed along a vertical line (Fig. 44). Such arrays direct the radiated sound into the horizontal planes which encompass the audience (for example, in an auditorium), and, because of the narrowness of the horizontal dimension of such arrays, they spread the sound in the horizontal direction so as again to encompass the entire audience.

It could thus be said that the disciplines of radar and electronic music became enmeshed, enabling an early development in radar to correct our intuitive impressions on beams and to provide us, thereby, with an enhancement in our high-fidelity sound reproduction.

MODERN ELECTRONIC MUSIC

We close this chapter with a brief mention of the most recent electronic music, best exemplified by the contributions of a now well-known and very outstanding innovator, Robert Moog (rhymes with *vogue*), musician, physicist, and inventor, who makes his Moog Synthesizers in Trumansburg, New York. There are now hundreds of his spectacular instruments in the United States, with many in colleges, where electronic music is taught to thousands of modern music technology enthusiasts.

A recent Associated Press account began: "Bells, hisses, bangs, rips . . . shrieks, gurgles, clattering oil drums, a pounding locomotive . . . symphonic flutes, horns, strings, and spine-shaking bass. All are electronic imitations, synthetic sounds, created entirely by a Moog Synthesizer." Other excerpts from that account note that movie sound tracks made by the Synthesizer include thundering climaxes which shake huge auditoriums with blasts comparable to a rocket taking off for the moon, and that because one man does the work of a 15-piece orchestra, the Musician's Union is concerned. The account continues:

> Even if they would demand that the wages paid must equal those of 15 musicians it wouldn't solve the problem of unemployment of the 15 men

not working, so they are asking "should we increase our membership rolls to embrace players of electronic instruments? Engineers?"

Today's electronic composers need a knowledge of electronic engineering as well as musicianship. They talk of signal paths, program sequences, percussive rhythms, melodic structures, transient peaks, tape delays, and sine waves, and manipulate oscillators, filters, amplifiers, noise reducers, reverb units, oscilloscopes and ring modulators, all part of an electronics explosion, an acoustical revolution, having tremendous implications for the future.

So electronic music has grown from the radio set squeal and the Theremin to electronic organs, electronic doorbells, and the Moog Synthesizer. Today, electronic music is a multimillion dollar business, and one can wonder how Heinrich Hertz and Lee DeForest would feel if they knew that their radio wave and radio tube discoveries are now benefiting music lovers the world over.

4

LASERS

Whereas in the last chapter we discussed the field of electronic music, a field that was initiated and became fairly well developed *before* the transistor came into being, we now examine a scientific discovery that *followed* the transistor by a decade or more, the laser. It is a device which, like the transistor, demonstrates quite markedly the value to an innovator of a broad, interdisciplinary knowledge. To illustrate this, we shall describe, following a brief introduction to the laser itself, some of the many applications of the laser in *nonphysics* fields. These are of interest because the laser itself required of its discoverers a very extensive knowledge of the physics of atoms and molecules.

THE LASER

To acquire a picture of the light from a laser, imagine a flashlight generating a beam of light which remains *very* narrow, not for just a short distance, like the standard flashlight beam, but for extremely long distances. With such a narrow "collimated," *straight* beam, the light energy remains concentrated within the beam and can accordingly be seen from great distances away. Figure 45 shows how beams of two lasers can travel from the earth to the moon and still display their brightness. A laser also generates very single-color light, a property which enables the light to be focused and thus concentrated into extremely tiny areas. This property is seen in Fig. 46, where a razor blade is having a hole punched in it by the focused light of a laser. (This figure was used on the jacket of my 1975 Plenum book *Engineering*

Fig. 45. Partially illuminated by the sun off to the right, the earth appears as a crescent to a U.S. moon probe, NASA's Surveyor VII. Directly above the arrow are two tiny spots of light formed by laser beams aimed toward the moon and originating at Kitt Peak, Arizona, and Table Mountain, California. Courtesy NASA.

Applications of Lasers and Holography.) Finally, recently developed lasers, some of which generate infrared (invisible) light waves, are extremely *powerful*. This is seen in Fig. 47, where the narrow, straight-line beam of an infrared laser is so powerful that it causes electrical breakdown of air through which the energy is traveling.

We shall see that it is these properties—the straight line beam, the single-color (single-frequency) property, and the ability to focus its high power into small areas—which have led to many of the laser's interdisciplinary applications.

LASER FUNDAMENTALS

Laser action is due to a physical process called "stimulated emission," and the name "laser" is derived from the first letters of the

Fig. 46. A steel razor blade disintegrates when the coherent light waves from a ruby laser are focused sharply on it. Courtesy Bendix Aerospace Systems Division.

Fig. 47. Even without focusing, the beams from present-day truly powerful lasers can create rather spectacular effects. Courtesy Hadron, Inc.

words *L*ight *A*mplification by the *S*timulated *E*mission of *R*adiation. Atoms which have been excited (that is, which have acquired energy from an outside source of excitation) were found, by the laser's discoverers, to be capable of releasing that extra (excited) energy in a very special way, one which would cause the released energy to have a single frequency (to have an extremely narrow-frequency bandwidth). The earliest stimulated-emission device to be developed was the microwave maser; it was demonstrated in 1954.* For its development, the Nobel Prize in Physics was awarded jointly in 1964 to the U.S. scientist Charles H. Townes, then at the Massachusetts Institute of Technology in Cambridge, Massachusetts, and to the Soviet scientists A. M. Prokhorov and N. Basov, both at the Lebedev Institute in Moscow. (The photo in Appendix 11 of Prokhorov and me, was taken in 1959 in Prokhorov's laboratory at the Lebedev Institute.) Following the success of the maser, many workers endeavored to extend its use from microwaves to light wavelengths. In 1960, the U.S. scientist T. H. Maiman, then at the Research Laboratories of the Hughes Aircraft Company in California, demonstrated the first laser, using a ruby rod as the active element.† His original laser is shown in Fig. 48. Let us examine the workings of this first laser.

* J. P. Gordon, H. J. Zeiger, and C. H. Townes, "The maser—New type of amplifier, frequency standard, and spectrometer," *Phys. Rev., 99,* pp. 1264–1274 (1955).
† T. H. Maiman, "Stimulated optical radiation in ruby," *Nature, 187,* pp. 493–494 (1960).

The active material, ruby, was shaped into a cylindrical rod. As shown in Fig. 49, a helical flash tube was wrapped around this rod, and when it was connected to a powerful source of stored electrical energy the flash tube emitted a very short and very intense burst of broad-band (incoherent) light. Some of this light energy was absorbed by the atoms of the ruby rod, and in this process the atoms were excited, that is, they were placed in an *energy state* at a higher *energy level* than that in which most of the atoms reside. Energy was thereby stored in these atoms, and when they returned to their normal *unexcited* state (we call

Fig. 48. The first laser. It was designed by T. H. Maiman at the Hughes Aircraft Company.

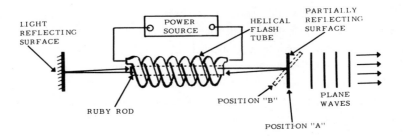

Fig. 49.   The essential parts of Maiman's ruby laser.

this the "ground state") they released their stored energy in the form of light waves.

Up to this point there is nothing special about the light-emitting process just described; practically all light sources generate their light in this way. Thus the atoms in an incandescent lamp filament are excited to higher energy states by the heat energy provided by the electric current passing through the filament, and as these atoms return to lower energy states (only to be again excited) they release this energy difference in the form of light.

The Danish scientist Niels Bohr suggested many years ago that the radiation of spectral lines by atoms could be explained by assuming that electrons revolve about the atom's nucleus in certain fixed orbits (like the planets circle the sun) and that each of these orbits represents a definite energy level. When an electron is in an outer orbit the atom is in a state of higher energy, an excited state, and when the electron transfers to an inner orbit, energy is radiated in the form of spectral lines that are characteristic of a particular atom. Through this theory, Bohr obtained values for the spectral series of visible hydrogen lines (colors) with an accuracy which was quite astonishing.

THE METASTABLE STATE

The significant difference between the laser and other light sources is that the laser light source materials provide a particular form of energy state in which the excited atoms can and do pause before returning to their ground state. They tend to remain in this state (called a "metastable" state) until stimulated into returning to the ground state.

In this last step, they emit light having exactly the same wavelength (the same color) as the light which triggered them into leaving that state. The atoms are thus stimulated into emitting, hence the words "stimulated emission" in the laser acronym. Stated another way, energy is first stored in the atom and later released by it; when it transfers from the metastable state to the lower one, the released energy is in the form of single-wavelength light energy.

THE TWO-STEP PROCESS

In the laser process, such as in Maiman's ruby rod, the wavelength of the emitted light waves is exactly the same as the wavelength of the light which stimulated the atom to emit; these new waves are thus exactly suited to react with other metastable atoms in the rod so as to cause them to emit more of this same radiation. One such laser process involves two steps, as shown in the energy diagram of Fig. 50. Such energy diagrams, representing atomic processes, are plotted with the lower energy states placed low on the vertical scale (in parallel with the situation where objects at lower heights have lower potential energy). The lowest level, the ground state, represents that energy level to which atoms in the metastable state transfer. The excitation energy of the flash tube imparts to atoms residing in the lowest level sufficient energy to raise them to the energy state represented by the highest of the three levels shown. From this level they fall to the middle level state, the

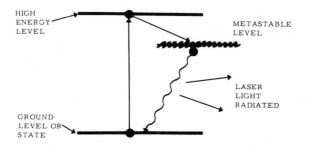

Fig. 50. Three energy levels are available to atoms or ions in a laser material; transition from the middle, metastable state to the ground state results in the emission of laser light.

metastable state, with the nature of this metastable level being such that the atoms tend to remain in it. When, however, the exactly correct wavelength light impinges on an atom in this state, it will depart from this state and fall to the lower energy state, emitting in the process a burst of energy in the form of light. The light burst which accompanies its return to the lower level can cause further emission of this same-wavelength radiation from other atoms which are residing in the metastable state.

We see, therefore, that laser action depends on the existence of this special state in the laser material. The process of placing, by means of the high-energy flash lamp, a large percentage of atoms in the laser material in the metastable state is referred to as a *population inversion* because originally most of the atoms reside in lower energy states. After the flash, an increased number of atoms reside in the metastable state.

GAS LASERS

A second and very important form of laser to be developed was the gas laser.* It consists most simply of a glass tube filled with a special gas mixture. In one form, a fairly high voltage is applied across two electrodes near the ends of the tube, as shown in Fig. 51, causing an electrical discharge to take place. The gas glows and the tube looks much like the glass tube of an ordinary neon advertising sign.

However, the gas laser differs from a neon sign in that its gas mixture provides the necessary metastable state in which the excited atoms can temporarily reside. As in the ruby laser, the energy difference between the metastable state and the lower state to which the atoms fall corresponds to the energy of the single-color light which is radiated. The first gas lasers used a mixture of helium and neon, and, many, like the first ruby laser, generated red light; their light had, however, an orange red color rather than the crimson red of the ruby laser. Other gases are now also used, for example, argon providing blue laser light and carbon dioxide providing invisible infrared laser light.

---

* A. Javan, W. R. Bennett, Jr., and D. R. Herriot, "Population inversion and continuous optical maser oscillation in a gas discharge containing a He-Ne mixture," *Phys. Rev. Letters, 6,* pp. 106–110 (1961).

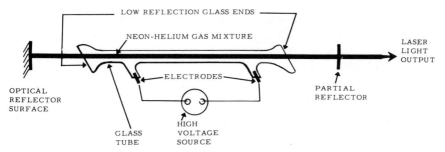

Fig. 51. Gas lasers often use a direct-current glow discharge for exciting the atoms or ions to the required higher energy levels.

One difference between gas lasers and ruby lasers is that most gas lasers operate continuously. The glow discharge caused by the applied voltage continually places a large population of atoms in a metastable state, and although many atoms are constantly falling back to a lower level, many are continually being elevated again by the glow discharge phenomenon.

Having reviewed some features of lasers in this brief introduction, let us move on to a discussion of laser applications.

SEMICONDUCTOR LASERS

Another form of laser is the semiconductor laser. It is similar to the ruby and gas lasers, except that it, like the transistor, employs as its active substance a tiny piece of semiconductor material. Figure 52 shows how tiny such semiconductors can now be made. Lasers have found numerous applications in the measurement field. Let us first examine some of the uses involving alignment and structural measurement.

ALIGNMENT

Because of the extremely straight-line nature of a tiny beam of laser light, systems using lasers have found wide use in surveying and also in industrial alignment applications. Such alignment systems have achieved accuracies 5 to 10 times higher than those of previous systems

Fig. 52. A tiny semiconductor laser placed on a U.S. coin. Courtesy Bell Telephone Laboratories.

using alighment telescopes. The arrangement shown in Fig. 53 has been in use at Boeing aircraft plants. In the photo, the laser is shown at the far right and the optical receiver at the far left. One of the structures of the Boeing 747 aircraft is being monitored during assembly in this arrangement. One aircraft builder sees the use of such laser systems as

Fig. 53. A typical laser precision alignment system, here used on a Boeing 747 wing skin jig.

inevitable because the size of today's (and future) aircraft requires assembly tools up to 200 feet in length and an alignment tolerance of 0.01 inch. In California, a laser was used to position dredges in San Francisco Bay during work on a subway tunnel. In 1973, a patent was issued on the idea of using a laser for measurement in football games. In this system, a battery-operated laser measures the distance a football team needs for a first down. The laser unit is in the forward pole marker, directing its beam across the field; if the football is in the beam's path, a reflection is observed, showing that a first down was achieved. This new concept was described in my 1975 book on lasers and holography because I considered it a rather clever idea. I was pleased to see a version of the device used and broadly described in the nationally televised (1976) North-South Senior Bowl football game.

DISTANCE AND LENGTH MEASUREMENTS

In echo-location systems such as radars (which employ radio-wavelength electromagnetic waves), the use of lasers provides certain advantages.* Thus the excellent detail possible in a (scanning) radar

---

* A. L. Hammond, "Laser ranging: Measuring the moon's distance," *Science, 170,* pp. 1289–1920 (1970).

record when a laser is employed as the radar transmitter is indicated in Fig. 54. The figure shows a laser radar record (left) of the airport control tower shown at the right. Another important military use of lasers involves the illumination of a target from an aircraft and an air-dropped bomb which contains an optical seeker. As the bomb descends, it "homes" on the laser-illuminated target, permitting the aircraft with its laser illuminator to remain at a safe distance away. These devices are referred to as "laser-guided bombs."

LASERS IN MACHINE TOOL MEASUREMENTS

Lasers have been used in a variety of machine tool applications, as an integral part of machine tool positioning and of control systems for sensing and for correcting machine tool geometry and also in precision inspection applications.* In the laser interferometer gauge, two beams are combined to form constructive and destructive interference regions (light wave "fringes") by virtue of a moving mirror. Sensors count the number of laser fringes observed as a function of the displacement of a movable stylus, thereby providing a highly accurate length measurement, without the usual requirement of so-called gauge blocks or masters (blocks having precisely established lengths).

INTERFEROMETRIC DETECTION OF FOOTPRINTS

Most composite or fibrous materials, such as wood, plastics, and woven material, exhibit the phenomenon of *creep* after being deformed. Thus a footprint deforms in a very minute way many floor surfaces, including rugs or carpets. When the foot is removed, the fibers of the material do not immediately return to their original shape, but creep back slowly, sometimes taking hours to do so. Such movements, however, are detectable with a technique called "interferometry." A photographic laser record is first made of the area where the

* W. E. Kock, "The use of lasers in machine tools," *Optical Spectra* (March 1970). Presented at the International Laser Colloquium, Paris, France (November 1969). Also in French: W. E. Kock, "L'emploi des lasers dans les equiments de mesure," *Mesures*, *35*, pp. 75–80 (April 1970).

Fig. 54. A laser radar picture, left, of an airport control tower "painted" from 14 sweeps of the laser radar beam. Courtesy United Aircraft Research Laboratories.

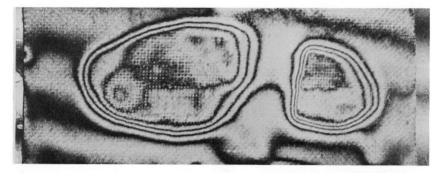

Fig. 55.   The outline of a footprint on a carpet, although not visible to the naked eye, is revealed in a reconstructed laser interference pattern.   Courtesy Hadron-Korad.

impressions are suspected, and some minutes later a second record is made on the same photographic plate. Differences in the reflected light resulting from the minute movements in the suspected area appear as fringes. Figure 55 shows a typical photo made by this technique. Scotland Yard has been examining the possibilities of using this technique in apprehending intruders.

LASERS IN THE SUPERMARKET

Perhaps the best-known laser application of recent months is the use in automating the checkout of items purchased by customers in supermarkets. This development, expected to be at the checkout counters in most markets by 1980, involves the laser "reading" of a code stamped on packages and the transmission of the information to computers for rapid determination of the total bill (Fig. 56). The computers tabulate the data into inventory information, prices, and sales taxes, and quickly give the customer the totals, the amount of his change, and the number of trading stamps. The checkout attendant, freed of cash register duties, can immediately go about placing the merchandise in bags. Savings, according to representatives of the U.S. grocery associations, can be quite high.

One of the first experiments in testing this development was conducted at a Kroger supermarket at the Kenwood Plaza Shopping Center in Cincinnati, Ohio, starting in the summer of 1972. This store has annual sales of \$3.7 million. In the checkout system used there (developed by RCA), a low-power laser beam is expanded in an optical system and caused to strike a rotating multifaceted mirror. The mirror scans the laser beam across a slot in the checkout counter and thus scans the universal (laser) code now appearing on practically all packages (Fig. 57) as they are slid over the slot. The reflected beam conveys the information to a detector, which then feeds it to the computer for processing. The price is recorded and displayed at a window facing the customer (Fig. 56). The test has been a striking success, according to Kroger's director in industrial engineering, Robert L. Cottrell, who noted that more than 3 million items were scanned without a mistake. More recently, IBM, National Semiconductor, and other companies introduced similar checkout systems. It has been predicted* that sales of the laser readers alone in such supermarket devices should in 1977 double the estimated \$1 million in 1976.

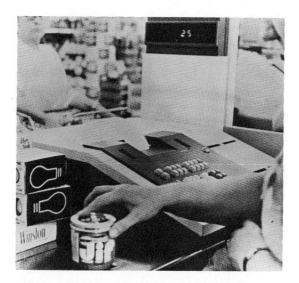

Fig. 56. At a Kroger store in Cincinnati, a clerk passes a jar across a scanner while a customer empties her cart. The price is computed and displayed automatically (top).

---

* *Laser Focus,* p. 34 (January 1976).

Fig. 57. This cartoon displays some samples of the universal laser code now found on packages sold in supermarkets. The codes permit the laser "reading device" to identify to the computer the item being purchased.

DRILLING AND WELDING APPLICATIONS

The remarkable controllability of the transverse physical size (diameter) of a laser pulse has made the laser useful in such precise applications as the balancing of the tiny balance wheel of watches and clocks by "drilling" tiny holes in it. Because a laser beam can be focused, even through clear glass or plastic, into a tiny, very hot spot, lasers are now extensively used in specialized on-site welding, permitting simple bonding and the joining of dissimilar materials and thermally incompatible work pieces, even in inaccessible locations (such as inside a glass envelope). The value of laser welding systems is greatest in tasks which defy conventional methods, such as in the welding of dissimilar metals and precision metals and alloys, in cases requiring a narrow heat-affected zone, and in the microwelding of extremely tiny parts.

IDENTIFICATION

One feature of the laser that has made it useful in law enforcement applications is its sharp focusing property, which permits extremely tiny but still identifiable grids to be marked on metal objects. Figure 58 shows how guns have been marked with an identifiable, indelible grid so that if the gun is stolen its reappearance at a pawnbroker's shop can give a lead to the thief. For permanent marking of serial numbers or other information, the Hadron Corporation has introduced a pulsed laser system that traces out as many as 30 dot-matrix characters per second.*

FABRIC CUTTING

Interest in the use of lasers for cutting fabrics has grown rapidly in the last few years, with numerous apparel manufacturers purchasing cutters, each costing over $500,000. The advantages are the ability to respond quickly to fashion changes, the better-fitting clothes which result, and the reduction of waste by cutting to very close tolerances without error.

MEDICAL APPLICATIONS

Promising medical uses of lasers are found in such fields as ophthalmology (cataracts, retinal detachments), dentistry (enamel glaz-

Fig. 58. Guns are marked with an indelible laser grid for identification.

* *Ibid.*

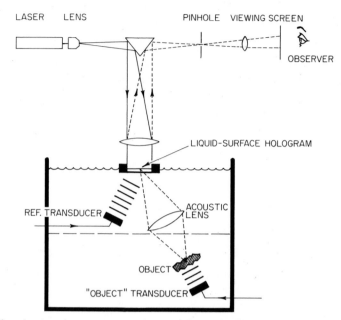

Fig. 59. An interference pattern is created on the liquid surface by two sets of ultrasound waves. An image of the object can be generated by illuminating the surface with a laser.

ing, tooth contouring), and laser beam surgery. Significant medical applications are also found in the field of "liquid surface" ultrasonic imaging (a form of acoustic "X-rays") in medical diagnostics. Figure 59 shows the principles of an instrument which is now available commercially and is used in numerous hospitals. In such liquid surface systems, no photographic development is required, and, because the liquid surface responds rapidly to the ultrasound energy, an instantaneous laser readout, even of a continuously varying picture, is available. If, in Fig. 59, a person's hand is placed where the "object" is located, a continuous, X-ray-like picture of the hand is observed, even as it is moved about (Fig. 60).

Figure 61 provides a comparison between the acoustic image and the normal X-ray. In this figure, the acoustic image is portrayed as a negative so that in both of these photos the bones are white. The ability to diagnose small breast tumors and cysts may prove to be the greatest service of this form of imaging. This capability has been convincingly demonstrated, and the prospects that extremely small tumors will be detectable appear excellent. Weiss has pointed out* that small lumps (1 millimeter or so in diameter) are not generally susceptible to detection by manual palpation. Yet if malignant lumps are not diagnosed at their early stages, the cancer cells spread to other parts of the body to form secondary growths not in contact with the primary cancer; this generally develops into "advanced cancer," which is virtually incurable.

As we noted earlier, lasers are used in surgery† and in dental labo-

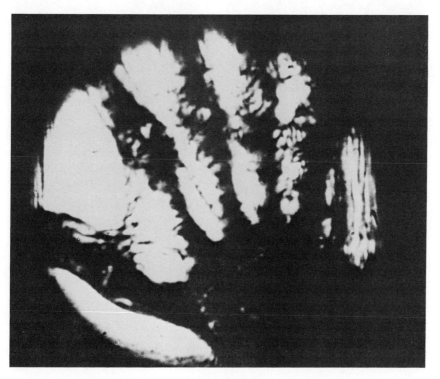

Fig. 60.   The acoustic image of a person's hand.

* L. Weiss, "Detection of breast cancer," in *Ultrasonic Imaging and Holography,* ed. by G. W. Stroke, W. E. Kock, Y. Kikuchi, and J. Tsuijiuchi (New York: Plenum Press, 1974), pp. 567–585.

† L. Goldman, R. J. Rockwell, Jr., Gunther Nath, and George Schidler, "Recent advances in laser surgery using Nd-YAG," Fourth Electro-optics Conference (September 1972).

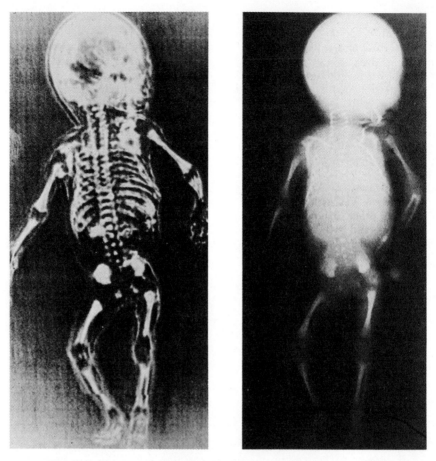

Fig. 61.   Comparison of an X-ray (right) and an ultrasonic image (left).

ratories.* A laser scalpel has been developed that is based on an infrared laser and a fiber optic delivery arm.† Initial applications of this instrument will be surgery of the ear, nose, and throat. Figure 62 shows a laser being used in surgery. Even the ancient Chinese art of acupuncture has bowed to the laser. A West German firm has introduced a laser device that is claimed to be superior to the standard needles of acupunc-

---

* T. E. Gordon, Jr., and D. L. Smith, "A laser in the dental lab," *Laser Focus,* pp. 37–39 (June 1970).
† *Laser Focus,* p. 35 (January 1976).

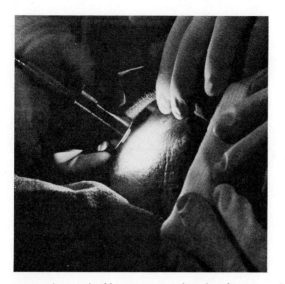

Fig. 62. A laser operating at the blue-green wavelengths of argon and neodymium is used at the Laser Laboratory of the University of Cincinnati to remove a reddish birthmark from a patient's face. The argon laser is ideal for destruction of pigmented tumors because its light beam is absorbed by red. From L. Wingerson, The Sciences, pp. 27–29 (August–September, 1975).

ture. In this equipment, the length of the time that a beam is applied to an acupuncture point is precisely controllable.*

COMMUNICATING WITH LIGHT

The strong demand for increased communications bandwidth has been continuously evident for many decades. From telegraph, to telephone lines, to coaxial cable, to microwave radio relay circuits, and recently to satellites, the capacity of communication circuits in frequency bandwidth has steadily been enlarged. The advent of the laser with its extremely high light wave frequency brought into being a new communications possibility, the concept of communicating with light. Already several short-range, low-powered, optical communications systems are available both in this country and abroad. Thus a light-

*"Laser allows needleless acupuncture," *Electronics*, p. 55 (January 8, 1976).

Fig. 63. An invisible CO_2 laser beam melts a glass rod, permitting the glass to be drawn into a fine fiber. The drum at the right holds the drawn fiber. Courtesy Bell Laboratories.

weight, hand-held laser communications system was recently used in demonstrations with the Los Angeles Police Department. Police envision using the unit in helicopter patrols that send and receive messages from ground patrol cars.

The most extensive use of lasers in communications could come as a technique for increasing the capabilities of the huge telephone networks both here and abroad. Extensive research is now under way to develop extremely fine glass fibers to act as "cables" or "waveguides" for transmitting speech, network television, and computer instructions.

Until fairly recently, such "wires of glass" exhibited too great a loss per mile to be acceptable. But now several companies, including the Corning Glass and Bell Telephone Laboratories in this country and the Siemens Laboratories in Munich, Germany, have developed such extremely low-cost fibers that Bell Laboratories, General Telephone (GTE), and the British Post Office will shortly initiate separate field trials of fiber optic phone systems.*

Figure 63 shows the use of a powerful carbon dioxide laser to melt a glass rod, thereby permitting it to be drawn into an extremely fine fiber.

* *Electronics,* pp. 25 and 53 (September 18, 1975).

VIDEO DISKS

An optical system that uses a laser beam for playback is the basis of a "videodisk" player developed for the consumer market by N. V. Phillips, the Dutch electronics giant, and MCA Inc., the U.S. producer of movies, TV, and records. The two developers wish to persuade other companies to manufacture and market the player for the U.S. audience under licensing agreements. In the Phillips-MCA system, the player features a 1-milliwatt laser and a tiny lens to focus its beam. The end result is a player that could be attached to any home TV set to play prerecorded videodisks that might retail for $2 to $10 for a half-hour show.*

LASER FUSION

One of the more long-range but still very exciting uses of high-power lasers is in the nuclear field. In this application, lasers may well play a key role in our future energy problems. Laser-induced nuclear fusion now appears very close to solving the world's oil, gas, and coal depletion problems, since it would utilize a form of hydrogen that abounds in sea water (estimated to be sufficient to provide for the world's energy needs for 350 million years). The first atomic bombs used the fission process, whereby the atoms of the element uranium were *split,* thereby releasing large amounts of nuclear energy. It is this fission process which is used in a controlled way in present-day nuclear reactors, now generating electric power in many nations of the world. Because the world's supply of uranium is limited (as is oil and coal), much effort has been applied recently toward "taming" the process utilized in the later hydrogen bomb. This process is called "fusion" because in it hydrogen atoms are fused together to form heavier atoms (helium). If, as in fission, the fusion process can be converted from the uncontrolled one used in the bomb to a controlled one, a nuclear power plant using it will have three highly attractive features. First, as noted, the "fuel" supply (heavy hydrogen) is practically unlimited; second, the process is a very "clean" one as compared to fission (there is no radio-

---

* "Videodisks with Lasers," *Optical Spectra,* p. 18 (May 1975).

active spent fuel to be disposed of); and, third, in the newest procedure (using lasers) there is no danger of an "accident," about which there is concern today in fission reactors.

In May 1968, Nikolai Basov (see Appendix 12) reported on experiments conducted at the Lebedev Institute in Moscow. He employed a laser having five amplifier stages generating an extremely powerful laser pulse that was directed against a nuclear target. He noted that in about every sixth experiment, neutrons were detected (an indication of fusion action). Figure 64 shows a chain of "amplifiers" which can increase the energy in the original laser pulse by a huge factor. An announcement which attracted worldwide attention came in 1970 from the French Atomic Energy Laboratories at Limeil (the Limeil Weapons Research Center), located just outside Paris. A joint effort between that organization and the French firm Compagnie Generale Electrique (CGE) has led to the development of an especially

Fig. 64. A chain of high-power laser amplifiers. Courtesy Hadron, Inc.

pure form of neodymium glass for use in laser rods. Using this glass, CGE was able to construct a laser amplifier chain that could generate a pulse much more powerful than the one Basov employed. When this laser was used to bombard a cylinder of frozen heavy hydrogen, approximately 1000 neutrons were generated with each pulse (Fig. 65). The energy of the neutrons observed was determined by means of a time-of-flight travel measurement over a known distance; the observed time of flight confirmed the assumption that the hoped-for fusion reaction had indeed occurred.*

The Limeil announcement made headlines in science journals the world over, with Fig. 65 appearing on the cover of several technical journals. Nevertheless, many remained skeptical of achieving, within any reasonable time, the ultimate goal, the generation of a significantly larger amount of nuclear power than that used to power the laser.

In April 1972, Keith Bruckner, of KMS Industries at Ann Arbor, Michigan, described a new way of shaping the laser pulse so as to cause a fusion "fuel pellet" to be *imploded* (highly compressed). Implosion increases the density of the pellet, and this provides a significant advantage because the desired nuclear reaction proceeds as the square of the density of the pellet. Experiments at KMS in 1974 showed visible volume compressions of 210 times, with actual compressions presumably much higher.†

With the formation of the U.S. Energy Research and Development Administration, the possibilities of laser-induced fusion were seen as quite significant, and ERDA announced a 55% increase (to $73.5 million) in the 1976 spending on laser fusion research. As one report noted, "When and if laser fusion is shown to be feasible, around 1978 according to ERDA's timetable, equipment purchases will begin a long and dramatic rise."‡

INTERDISCIPLINARY INNOVATION

The world owes a debt of gratitude to those medical scientists, to those radar and communications engineers, and to those industry

---

* William D. Metz, "Laser fusion," *Science,* pp. 1193–1195 (December 27, 1974).
† *Ibid.*
‡ *Laser Focus,* pp. 16 and 18 (January 1976).

Fig. 65. A rod of frozen heavy hydrogen becomes, when a focused laser beam of 4 billion watts strikes it, a tiny hydrogen bomb, thereby holding promise for an energy source using the plentiful water of the oceans as a "fuel." Courtesy Hadron, Inc.

experts who, through their broad backgrounds of knowledge, were able to understand sufficiently well the physics of the laser and its characteristics to be able to recognize not only those applications possibilities just described but many others of equal importance. Without such innovators, the laser might have remained for many years a laboratory curiosity, and the beneficial economic impact on the nation and the world of its wide fields of application might have been greatly delayed.

The transistor was never threatened by such a possible fate of seclusion, because it was immediately seen as the perfect replacement for the radio tube amplifier. But the laser's fate could have been quite different because it was the very first of its kind. During the first few years of its life, it was often referred to as a solution looking for a problem. However, a recent study* forecasts that laser sales will quadruple by 1979, creating a $4 billion market for systems manufacturers, aerospace firms, and suppliers of laser components and accessories.

* *Electronic News,* p. 40 (November 17, 1975).

5

TRAITS

Having reviewed three important engineering innovations of the last half century or so, let us now discuss various *traits* of innovators and of those who help others, such as their children, to *become* invention conscious.

HELP FOR THE YOUNG

The U.S. Department of Health, Education, and Welfare has, since 1954, been issuing a bulletin entitled *How Children Can Be Creative*. The opening paragraphs (1960) note that

> As children become acquainted with their world they begin to use the objects and materials around them to express their thoughts and feelings. Teachers and parents can do much to provide a stimulating environment in which children feel free to create.
>
> Who is the creative child? He is every child in every classroom and in every home. Some children create as a natural form of expression, without adult help or stimulation. Other children need an environment which supplies both materials and the encouragement of a parent or teacher. Children are creative in different degrees in different media. They may draw, paint, engage in dramatic play, express themselves through pantomime or impersonation, or with puppets or other dramatic forms; may develop rhythms, or dances, write music to be sung or played, or make a simple instrument; may model with clay, weave, cook, sew, build with blocks, and write stories or poems. These and other related forms of expression are natural to many children. Many of these experiences when carried on at school may be drawn together in such a way as to supplement each other, and so enrich the whole school program.

Because each child is an individual, he may prefer one medium to another. A function of the adult is to help him get at least some experience with as many of these as possible. Teachers and parents need to develop an attitude of mind which will permit children to experiment freely in any medium that will encourage the desire to create.

Creativeness calls for a willingness to experiment, to be independent, to express original ideas without regard to how others may feel about them. In creative expression, the child attempts to show what he feels about people, objects, or an experience. He is not concerned with reproducing what someone else has done. Only one who has known the thrill and the uplift of creating something that is his own can fully appreciate the importance of this experience to the individual. Children as well as adults need continued encouragement to express such ideas in many different forms.

THE VALUE OF PRAISE AND ENCOURAGEMENT

The HEW bulletin's reference to the creative person's need for *encouragement* points out one of the most important assists which parents can give to their children in guiding them on their road to creativity. Thus in the Bulletin's section entitled "What Is the Role of Parents in the Creative Expression of Children?" it is stated that

> it is important that what the parent says about a child's creative effort should encourage the boy or girl to continue to create with pencil and paper, with crayon or paint, or with a handicraft. If a parent says of a child's picture which he has brought home, "Do you call that a house?" or "You'll have to learn how to draw people," by so doing he may discourage further effort on the child's part. The boy or girl may not bother to bring his next picture home for the family to see. Later he will leave his piece of pottery, his wood carving, or piece of weaving at school, rather than risk adult criticism. But if, in contrast, the parent says, "Tell me a story about your picture," he encourages the child to make further creative efforts, as well as to give him something to talk about that is of real interest.
>
> Children are not adults. Therefore their work can be judged only in terms of what each individual child is capable of doing. The self-portrait which a 6-year-old makes may have no resemblance to him, but it is his best effort to transfer to paper his idea of what he looks like.
>
> Most important of all, parents should recognize the creative arts as a safety valve, as an emotional outlet for children that will contribute to their development, or will prevent difficulties from arising.
>
> That child is fortunate who finds in his home good examples of creative art, of music, of literature; who sees in the choice of colors, and of furniture and room arrangement the use of art principles in creative ways. The

arrangement of flowers, the setting of the table, the selection of clothing, the use of color in planning a garden can all give evidence that creative art is an important part of daily living.

THE VALUE OF CREATIVITY FOR CHILDREN

In its concluding section entitled "What Are the Major Values of Creative Experience for Children?" the HEW bulletin notes that

Creative experiences help the child develop as an individual in many ways. They provide him with varied ways of expressing his thoughts and emotions. He can communicate more easily and effectively with other children and adults than the child with limited means of expression.

From creative experiences comes enriched living for the child. He finds many avenues of enjoyment through creative experiences in music, arts, crafts, writing, drama, and other forms of expression. Some of these experiences involve creative listening or observing; others involve the actual creation of something, through individual or group effort. Children carry on some of these activities at school, some at home, and some in other parts of the community.

Through having a part in creative experiences the child has an opportunity to participate in a highly personalized activity. It gives him much personal satisfaction and a tangible feeling of accomplishment that can give him emotional security. Rich experiences in creative expression help to fulfill the goal of education labeled as self-realization.

To enable a child to enjoy these values, parents should keep uppermost in their minds the value of praise, encouragement, and understanding. The following quotations from the "Thoughts" section of *Forbes* magazine discuss praise and approval:

They that value not praise will never do anything worthy of praise.
—Thomas Fuller

I have yet to find a man, whatever his situation in life, who did not do better work and put forth greater effort under a spirit of approval than he ever would do under a spirit of criticism. —Charles M. Schwab

The thing most people want is genuine understanding. If you can understand the feelings and moods of another person, you have something fine to offer. —Paul Brock

Each of us is a little lonely, deep inside, and cries to be understood.
—Leo Rosten

We sometimes imagine we hate flattery, but we only hate the way we are flattered. —La Rochefoucauld

Mankind differ in their notions of happiness; but in my opinion he truly possesses it who lives in the anticipation of honest fame, and the glorious figure he shall make in the eyes of posterity.   —Pliny the Younger

The deep-down desire to stand well with our fellows, this laudable hunger to win the public's "well-done, good and faithful servant," is inherent in every normal human soul. We may flout it when we are young or even in our prime, but when we begin to cast up our life's reckoning, we cannot ignore it. We then realize that this is one of the things that count, one of the things really worth while, this esteem and good will of our brother mortals, and the knowledge that we have sincerely tried to earn it.

—Bertie Charles Forbes

Napoleon had similar words:

Happiness is the highest possible development of my talents.

And had these very wise words for the young:

He will not go far who knows from the first whither he is going.

And Ralph Waldo Emerson once said:

Nothing great was ever achieved without enthusiasm.

OTHER TRAITS OF CREATIVE YOUNGSTERS

Many children have thoughts of great accomplishments, and an interview of the parents of the 1960 winners* of the Nineteenth Science Talent Search conducted by Science Clubs of America (for the Westinghouse Science Scholarships) revealed the following conclusions:

1. Show confidence in your children's abilities and leave them alone to work out their projects.
2. Never underestimate what they might be able to accomplish.
3. Use infinite patience in listening and always show them your sincere interest in any topics they want to discuss.
4. Provide quiet time, alone.

Parents can also encourage their children by occasionally repeating to them various fairly well-known statements such as the one made by the American universalist Hosea Ballou (1771–1852):

Idleness is emptiness. The tree in which the sap is stagnant remains fruitless.

* *Science News Letter,* p. 362 (June 4, 1960).

An interesting reference to this was made in the well-known comic strip "The Born Loser" in which a person handing a coin to a beggar quotes the above, adding Ballou's name at the end. The beggar, after some consternation, shouts to the departing giver: "Sticks and stones may break my bones but names can never hurt me—Mother Goose!"

Other useful quotes are "*Not* to decide, *is* to decide," and "To avoid criticism, do nothing, say nothing, be nothing" (E. Hubbard), and "Don't be a cloud just because you can't be a star."

What, in my opinion, was one of the finest expressions of advice for young people was composed perhaps a decade ago by Dr. Elizabeth Wood of the Bell Telephone Laboratories (see Fig. 177), in her discussion entitled *Rewarding Careers for Women in Physics* (an American Institute of Physics Brochure). Since, in my view, the beauty of her advice applies almost as equally to creative *engineers* as it does to physicists, I have taken the liberty of substituting the word "engineers" wherever "physicists" appeared in her original closing paragraph and "engineering" wherever "physics" appeared:

> Engineers are well paid today, but you had better not go into engineering for that reason. Engineers are regarded with awe by many people and are front page news, but you had better not go into engineering for that reason. Engineers have an all-pervading curiosity about every physical detail of the world we live in. The theoretical engineers enjoy doing the elaborate mathematical puzzles that are called theoretical calculations. The experimental engineers enjoy trying to design their experiments in such a way that the results will give unambiguous answers to the questions they are asking of nature. To share in this enjoyment, to make some part of the world of engineering your own—that is the best reason for becoming an engineer.

INTEREST IN PROBLEMS

Possibly one of the most valuable traits that parents can imbue their children with for enhancing their creative talents is an interest in *problem* solving. We shall discuss later my interest, during my younger years, in the pastime of solving chess problems. With the many chess books now available (in addition to the many books on the *game* of chess), the encouragement, by parents, of the child's acquiring an interest in chess problem solving can be valuable in improving his creative capabilities.

Koestler has suggested that "the *common element* of creativity in humor, art, and science is the solution of a problem."* For example, when progress in a situation is blocked (using the standard approaches) such a solution often involves switching from one way of thinking (one "matrix of thought") to another that is normally not included in that first context. Thus Koestler believes that "creativity is brought about by the 'collision of matrices,' that is, when two different ways of thinking about a situation are deliberately focused on a problem and the conditions are ripe for an interchange." Blank suggests that this concept of Koestler's is "particularly important in a scientific world where activity is increasingly constrained" and notes that the concept "appears to have been utilized within the framework of normal science by the creation of new interdisciplinary areas of research."†

CHESS-PLAYING TRAITS

A very revealing discussion of the chess interests of two Nobel Prize winners in chemistry appeared recently:

> Chess and chemistry wouldn't seem to have much in common—except that they are neighboring subjects in the encyclopedia. A friend of mine speculated recently that many eminent chemists get hooked on chess when as youngsters they look up the encyclopedia article on chemistry and stumble onto the nearby article on chess.
>
> Whether anyone will accept that explanation or not, it is remarkable that two of Britain's greatest chemists have also been talented and enthusiastic chessplayers. Sir Robert Robinson, who died in February of 1975, was a winner of the Nobel Prize for chemistry in 1947 and so many other prizes and honors that they cannot even be listed here. His main fields were the chemistry of natural products and the application of electronic theory to organic chemistry. He was perhaps the best known for helping develop the chemistry of penicillin.
>
> As a chemistry professor at Oxford University, Sir Robert twice won the Oxfordshire chess championship, and in 1956–57 he finished fourth in the British correspondence championship. He was president of the British Chess Federation from 1950 to 1953.
>
> Upon Sir Robert's death, one of his former students at Oxford wrote about the great chemist as follows: "I was a student of Robert's at Oxford

---

* Arthur Koestler, *The Act of Creation* (New York: Dell, 1967).
† M. Blank, "Interdisciplinary approaches in science," *European Science News,* Issue 29-9, p. 397 (U.S. Office of Naval Research, London).

and we were collaborators and friends for 35 years, sharing interests in both chemistry and chess.

He had some of the qualities as a chessplayer that made him pre-eminent as a scientist: a superb intellect with imagination, depth and the ability to see all around a problem. It was impossible to be bored during a game with him!

We played little across the board during his later years, but combined our scientific correspondence with games of postal chess. When his sight failed a few years ago, we continued to play postal chess without use of boards or men. These games when played over seem of good quality and are an indication of the mental powers that Robert retained to the end."

This former student is John W. Cornforth, and he has himself just won the Nobel Prize for chemistry. He is widely recognized for his work on enzyme catalysts.

Cornforth is now a professor of chemical enzymology at Sussex University in Brighton, England, and, like his eminent teacher, has combined high excellence in both those subjects that live as neighbors in the encyclopedia.

Cornforth was originally from Australia. While still a teenager, he acquired a distinguished reputation as a player, especially after he set the Australian record for blindfold exhibitions in 1937. He played 12 games, simultaneously, winning 8, drawing 2 and losing 2.

The games from this exhibition were against strong players, according to the Australian sources, making Cornforth's achievements remarkable indeed. Several of these games would be jewels in anyone's collection of regular over-the-board games, much less in a blindfold simultaneous exhibition.*

During my early student years, I also found chess a very enjoyable exercise of the mind. My father had gotten me interested in the game at a fairly early age and at 18 I experienced the pleasure of playing— simultaneously and while I was blindfolded—four opponents who were fellow University of Cincinnati engineering students. A fellow member of the University's chess team, Ernst Theimer, wrote of this "exhibition" in a letter to Dr. P. G. Keeney, the chess editor of the *Cincinnati Times-Star* as follows:

CHESS ACTIVITIES AT VARSITY
  November 28, 1928.
Chess Editor, Cincinnati Times-Star.
  Dear Sir: It may be of interest to you, for the column, to learn that Winston Kock, in a simultaneous blindfold exhibition at the University on November 20, defeated three out of four oponents. The oponents were

---

* D. Thackrey (chess editor), *The Ann Arbor News*, p. 46 (November 16, 1975).

some of the strongest in the Engineering College and the exhibition lasted only one and one-half hours.

Dr. Keeney composed a fairly lengthy and humorous article about this blindfold performance, with its title referring to the famous legendary chess player, Paul Morphy:

SPIRIT OF MORPHY HAUNTS U.C.

For a long time vague and uncertain but persistent rumors have been carried by the already heavily burdened airplane and radio charged atmosphere to the ears of the Time-Star chess editor. That worthy, an unbeliever as regards the supernatural, refused to make public the nature of the rumors until such time as verification of the manifestations could be had from a reliable source. Now that a letter from such a source has come to hand, the chess editor is compelled to enlighten his gullible readers with the startling news that the corridors, classrooms, campus, etc., at the University of Cincinnati are haunted. Yes, actually haunted! A ghost—a spirit of the greatest chess player that ever lived—is walking, walking and playing chess with the students and professors who have knowledge of the game. Paul Morphy, listed as an immortal, has been dead a good many years, but his spirit still mingles with the devotees of the royal game and haunts places where followers of the game, at which he excelled, are gathered.

This spirit at U.C. does not give vent to fits of ghoulish glee nor does it terrorize the inmates of the institutions by rattling chains, beheading itself or disappearing through a convenient keyhole. It is not that kind of a spirit. It is the ghost of Morphy walking again with his spirit embodied in the person of one Winston Kock, talented and popular young U.C. student and chess player of considerable skill. The attached letter, received by the chess editor, is partially explanatory of the uncanny happenings at U.C. recently and will do more in revealing and unmasking the ghost than any further literary effort on our part.

Soon the crash of falling pawns rent the air. Move followed move. Black played after White, and still Kock made no blunders, but kept the location of all the sixteen times four pieces in mind upon four mental boards of sixty-four squares, all in his mind alone. In the words of Homer, or Edgar Guest:

. . . and till the wonder grew
That one small head could carry all he knew.

And he won three out of four. In vain did three stalwart pawn pushers try to lay bare Kock's kings and give check. Each time did BLINDFOLD Kock interpose before his king a BLIND MAN'S BUFFER.

But one of the novices trimmed the redoubtable Winston. We must record the name of J. W. Hargy, as yet unknown to fame in Caissaterra, this lad will bear watching. For Kock is no easy man to defeat, even when he plays sight unseen.

This was a memorable exhibition. The guileless college youth are still agog over the feat of arms. Kock has done credit to the U.C. Chess Club and the Cincinnati Chess Club; may he head the university team, of which he is captain, to victory over State, Miami, and the rest.

Can it be that from the galaxy of U.C. stars another Morphy is about to rise?

The Cincinnati chess team was successful in the tournament mentioned near the end of the article, against Ohio State, Miami University, the University of Dayton, and others (Appendix 13). During the following year, the Cincinnati team again defeated Ohio State, resulting in one (Cincinnati) newspaper account having the following paragraphs:

"Champion chess team of Southern Ohio" is the mythical title claimed by the U.C. chess experts following their sensational victory over Ohio State last Saturday night. The match was held at Columbus, the lair of the Big Ten school.

The match itself was a thrilling one with the result in doubt until the final move took place. When Winston Kock, captain of the U.C. team last year, checkmated Marceau of State in the final game of the evening, a checkup of the various matches revealed the fact that the University had won, five to three.

As a result of several sensational victories achieved during the past two years the Cincy team recently laid claim to the Southern Ohio championship. Word came down from Columbus that that institution also thought that they had an excellent right to the same championship. There was nothing left for the local wizards to do but to make a hurried trip to the opponents' stronghold and convince them that they had the better right to the claim.

The other feature was Winston Kock's startling double victory over Marceau who was probably the most wary and canny man on the up-stater's team. Great credit is due Kock for this victory. His triumph is all the more appreciable in view of the fact that it was achieved after it became known that this match would decide the contest.

The 1928 blindfold event was referred to as recently as 1965 in a writeup about my NASA appointment:

WINSTON KOCK: AT U.C., PLAYED CHESS BLINDFOLD NOW HEADS NASA'S NEW RESEARCH CENTER

One of the newest things in the space age is NASA's Electronics Research Center in Cambridge, Mass. At its head is U.C. alumnus Winston E. Kock, EE '32, MSS '33, DSc (Hon) '52. For Dr. Kock, it is the latest adventure in a life of brilliance and inventiveness.

Even when he was a student at U.C., Kock's energy and ingenuity were apparent. He was on the varsity track team, wrote music for his college shows—and once, blindfolded, took on four of his college acquaintances at chess, winning three of the games. For his thesis, he came up with the invention of the electronic organ, and he later spent some years in developing a commercial version of his organ for the Baldwin Piano Company in Cincinnati.*

CHESS PROBLEM COMPOSING

Interest in chess problems was mentioned earlier, and whereas problem *solving,* as noted by Koestler, is an important element in creativity, the *composing* of an outstanding chess problem, involving as it does the "creation" of a new entity, obviously requires creative thought comparable to that exhibited by composers, authors, and painters.

At the beginning of this chapter, we referred to the U.S. Health, Education, and Welfare bulletin which discussed the many different media in which children exhibit creativity, noting that "they may draw, paint, . . . write music to be sung or played, . . . and write stories or poems." Because chess is a game which exercises the mind and requires imagination, the encouragement of youngsters to try to adapt their chess-playing ability to that of chess problem composing is, in my opinion, a worthy endeavor for stimulating their creative abilities. Chess problem composing tourneys are held moderately often, with newly composed problems being submitted by problem composers from all over the world. Should the young composer win a place in the award list of such a tourney, this recognition of his creative ability is comparable to the recognition given to those who have had their first musical composition (such as an anthem or a song) accepted by a music publisher. Successful chess problem composing can thus inspire the youngster to experiment with other avenues for creative thought, such as music, art, and technical innovations.

I found both chess problem solving and chess problem composing very rewarding, particularly because these two uses of the game of chess require no "opponent," as a normal chess game does. The pastime can thus be enjoyed alone, such as during a trip by airplane. As the young chess problem solver finds solutions to the published problems, he soon realizes that each one involves a novel approach toward making the

---

* *Cincinnati Alumnus,* pp. 14–15 (July 1965).

problem an acceptable one, that is, one which is "difficult" and therefore not solved easily by a beginner. This recognition of innovation in a chess problem can inspire him to devise a new approach to difficulty, often enabling him to "invent" a new concept (a new theme) around which he can fashion (compose) a new chess problem. The final working out of that problem usually requires much diligence and mental effort in order not to end up with a problem having *two* solutions. As in music composing, his first efforts are usually not too world shaking, but with practice the young chess problem composer soon produces one or two problems which turn out to be quite acceptable to a chess journal or a chess column (newspaper) editor. With a published problem under his belt, his way, from then on, is made.

I had this experience at the age of 20. In an international "two-move problem" composing contest, announced by the *Cincinnati Times-Star* in the spring of 1929, and to which 90 (new) problems were submitted by composers from all over the world, including "prominent and famous composers of Hungary, Russia, Belgium, Poland, Germany, and Canada,"* the two problems that I submitted won the second and sixth prizes (Appendix 14). The chess editor (Dr. P. G. Keeney) had selected as judge, Maxwell Bukofzer, Director of the Problem Division of the National Chess Federation, and the following, referring to my second-prize problem (Fig. 66), appeared in the tourney announcement:

> Concerning this prize winner the judge writes:
> "Mr. Kock undubitably possesses what I like to call a 'chess mind.' Whatever chess minds create is certain to show traces, if not traits, of that rare and elusive quality termed originality.
> "No. 5 is a problem uniquely towering above the other 89 entries of this tourney. In conception, workmanship and theatrical effect, it is supreme. The noble theme is treated with boldness, adroitness and humor. The variations, like links of a chain, are coherent, uniform, companionable. The character of the problem is Loydesque. The dearth of pawns betters the appearance if not the substance of the problem. In one word it is an unusual masterpiece."

COMPUTER CHESS

In an article by L. A. Steen discussing the programming of the "artificial intelligence" of computers for playing chess, it was noted that

* *Cincinnati Enquirer* Section 4, p. 6 (January 19, 1930).

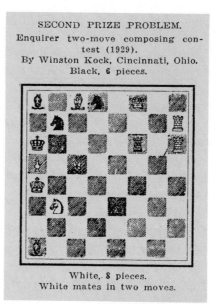

SECOND PRIZE PROBLEM.
Enquirer two-move composing contest (1929).
By Winston Kock, Cincinnati, Ohio.
Black, 6 pieces.

White, 8 pieces.
White mates in two moves.

Fig. 66.   One of the author's prize-winning chess problems.

Early pioneers in artificial intelligence—Claude Shannon, Alan Turing, Norbert Weiner—championed chess as an ideal proving ground for research because it posed a problem of sufficient definitiveness to permit assessment of the quality of the work being done, yet of sufficient complexity to prohibit a trivial solution based on the computational ability of the computer. Chess programs require heuristics—rules of thumb—that simulate the process of human thought; thus, chess programs form an excellent counterpoint in the study of human intelligence.

Shannon estimated that there are more than 10^{120} different possible sequences of moves from the beginning of a standard chess game. The fastest present computers take more than 10^{90} years to examine all these possibilities; even the fastest theoretically possible computer would take only slightly less time—10^{80} years! So whereas computers can "crunch" tic-tac-toe, and even checkers, by looking ahead all the way to the end of the game, they cannot do this with chess. That is why chess is such a challenge for players and for researchers in artificial intelligence.*

Another section points out:

Research into computer chess began a quarter century ago when information theorist Claude Shannon proposed a general method for constructing a

---

* L. A. Steen, "Computer chess: Mind versus machine," *Science News, 108,* p. 345 (November 29, 1975).

computer program to play legal and respectable chess. Research efforts since that time had led to about 40 moderately good programs around the world—about 25 of them in the United States alone.

But the game of chess is extremely complicated, and the article states:

Several of those closest to the computer chess effort predict that it might take as much as a half a century more before a machine's artificial intelligence can match chess wits evenly against the best natural intelligence.

The conclusions noted by Steen tend to confirm the concept that the game of chess is valuable as a means for improving the creative ability of the young.

MUSIC

The HEW bulletin referred to earlier suggested that the creative child may "write music to be sung or played." Parents often encourage their children to learn to play a musical instrument, with the piano being the instrument most often selected. This encouragement can be very valuable because the process of learning to play the piano involves a large amount of brain activity, and even though the young piano player may not reach great heights in this profession the exercise of his mind which is involved and the knowledge he acquires of the works of famous composers *can* lead him to try the process of *composing* music. As in the case of chess problem composing, the value of the music composed is determined by its acceptance by someone. In the beginning stages, the praise and encouragement of parents are very important. Later, his schoolteacher may choose one of the numbers for class use. The final mark (much later) occurs when a music publisher accepts it.

In my case, my parents encouraged me to take piano lessons at a very early age, often asking me, as I recall, to play one or two even very simple numbers for guests at their parties. I also recall while in grade school telling my parents that I did not want to continue piano lessons because my piano teacher, after I had practiced and learned, on my own, sections of Grieg's *Peer Gynt Suite,* told me that those selections were too advanced for me, that I should stick to simpler numbers. My parents did not say "you *must* continue," but rather agreed that my then teacher was not the proper one. A month or so later they suggested that I continue my lessons with another teacher, one who had initiated a high school piano course. This teacher, from the Cincinnati Conservatory of Music, Mrs. B. E. K. Evans, encouraged me to learn much more difficult numbers, so that as a high school freshman I participated in a concert (Appendix 15), and by the time I was a high school senior my repertoire was such that I was able to play, from memory, a full evening's recital (Appendix 16).

As we noted, one advantage of encouraging the child to learn to play a musical instrument is that this involvement could suggest to him at some point in time the idea of composing music. In many colleges, new original musical comedies are performed each year; these give student authors and music composers an opportunity to demonstrate their creative skills. Here again the excellence and usefulness of a musical composition are assessed, usually by an outside producer and his advisors. Acceptance by the producer is thus a recognition of talent, comparable to some extent to the acceptance by a publisher of a chess problem, a song, or a novel.

During my early college years, I had begun studying organ and was soon successful in securing a position as church organist and choir director in a small church. I found intriguing the idea of setting the words of simple anthems or hymns to new tunes, as I could have copies made and then foist them on my volunteer choir for a Sunday performance. Because the choir often liked some of these (and others they did not!), this experience helped me to learn what kind were acceptable. Appendix 17 was the most popular among my choirs. In my junior year at college, I decided to try my hand at composing tunes for the yearly musical comedy at the University of Cincinnati, and I was fortunate in having numerous selections accepted (as indicated in Appendix 18) and therefore published (Appendix 19). During my senior year, I also participated in that year's musical comedy production, and again was fortunate in having compositions accepted (Appendix 20 and 21). A 1931 newspaper article reported:

> among the compositions selected by Harry Wilsey for presentation by his orchestra at the second "Cincinnati Composer's Night" at the Zoo Clubhouse is "The More I Think of You" by Winston Kock, written for the 1931 University of Cincinnati Fresh Painter's show. Wilsey said Tuesday that "Winston Kock, in my opinion, is one of the cleverest amateur composers in the city."*

---

* *The Cincinnati Times-Star* (Tuesday, June 16, 1931).

Even during my year at the University of Berlin (1933–1934), I was asked to play a number of my own compositions at the Humboldt Club, in a joint piano recital (Appendix 22) called "Der Humboldt-Klub komponiert" (The Humboldt Club Composes). Some of my compositions were used in various churches in Cincinnati (Appendix 23).

RELIGION

We touch only briefly on a trait that some innovators encourage, the seeking of new ideas through religious readings.

I remember vividly how my father became quite excited over his discovery that certain words in the 22nd Psalm appeared to explain what was to him a biblical "mystery." My father had always felt that Jesus' words on the cross "My God, why hast Thou forsaken me," were not at all characteristic and he therefore kept wondering whether they were pointing to something. For him the following words, from the 22nd Psalm, were the answer:

Psalm 22

My God, My God, why hast Thou forsaken me? They pierced my hands and feet. They part my garments among them and cast lots upon my vesture. They say "He trusted in God, that He would deliver Him; let him deliver him if he would have him."

Later it continues:

The poor shall eat and be set free. They that seek after the Lord shall praise Him, your heart shall live forever.

My father was convinced that Jesus' words on the cross were telling the world of the much earlier description of that happening.

In another instance, Father Louis Dames of St. Patrick's Church in Governor's Harbour, Eleuthera, interpreted the following words of St. Paul as saying, in another way, those thoughts expressed in the first chapter about children having creative talent which is often lost as they grow older:

When I was a child I spake as a child, I understood as a child, I thought as a child; but when I became a man, I put away childish things. For now we see through a glass, darkly; but then, face to face. (I Corinthians 13:1)

My organ playing in church was probably responsible for my initiating, in 1932, the practice of preparing each year original

Christmas greetings (Christmas cards) which contained appropriate religious words, set to music. Appendix 24 is an early one (1934), and Appendix 25 (1940), 26 (1949), 27 (1971), and 28 (1975) are other examples. Again, comments from recipients indicated the acceptance of the composed music. The words chosen were always by others, with the very impressive ones in Appendix Item 28 having been heard by me while attending a 1974 Christmas Eve service at the Pleasant Ridge Presbyterian Church in Cincinnati (Clyde York, Pastor).

Finally, in religion the innovator often gets hope and optimism (for *new* inventions!) through such words as the following from the general Thanksgiving prayer: "We thank Thee . . . for all the blessings of this life . . . for the means of grace . . . and for the hope of glory."

As Warren Bennis (whom I quoted in Chapter 1) has noted:

> There is a beautiful interrelatedness in all the pursuits of man, whether it is of science, which is essentially the pursuit of truth; of aesthetics, the pursuit of beauty; or of ethics, the pursuit of goodness. John Keats was criticized for being redundant, or meaningless, when he wrote, "Beauty is truth, truth beauty," yet every great scientific truth has a beauty to it, and the greatest beauty of art often has a hidden scientific truth within it.
>
> The more our work makes us specialists, the more we must strive to become generalists in other matters, to perceive the interconnections between science, aesthetics, and ethics, to avoid becoming lopsided—like Darwin's gardener who said of him: "Poor man, he just stands and stares at a yellow flower for minutes at a time. He would be far better off with something to do."

Bennis also quotes the 8th Psalm:

> 4. What is man that thou art mindful of him? And the son of man that Thou visited him?
> 5. Thou madest him lower than the angels, to crown him with glory and worship.
> 6. Thou makest him to have dominion of the works of Thy hands; and Thou hast put all things in subjection under his feet.
> 7. All sheep and oxen; yea and the beasts of the field;
> 8. The fowls of the air and the fishes of the sea; and whatsoever walketh through the paths of the seas.
> 9. O Lord our Governor, how excellent is Thy Name in all the world.

AUTHORSHIP

Many engineering and science innovators have moved into the field of science fiction writing. Here too it is the novel ideas (innovative ones) which cause such books to sell. Two very successful such authors are Arthur C. Clarke (we shall hear more about him in Chapter 8, discussing satellites), and the well-known Bell Laboratories inventor John R. Pierce (responsible, with Rudolph Kompfner, for today's very important microwave amplifier, the "traveling-wave" tube). Pierce has written under the name of J. J. Coupling, after the *j-j* coupling expression in modern physics.

Another example of innovativeness in authorship is found in mystery novels, and one of the most renowned such authors, Agatha Christie, who wrote 84 mystery novels, received the honor of a page-one story in the *New York Times* (just a few months before her death). A mystery is effectively a *problem*, and those for which the *solution* as to "who done it" is easy are not rated very highly. In chess problems, the expression "tries" is extensively used; these are possible "first moves," which lead to checkmates for *many* of black's possible replies but not for *all* of them. The would-be solver thus imagines that each of these first moves (tries) is the true solution, only to find that black has a response by which he avoids the required (so-many-move) checkmate. A good murder mystery similarly has a number of (likely) suspects, and the reader-solver of a good mystery usually picks the wrong one.

Authors of cartoons and humorous columns also must come up with novel ideas if they are to maintain a following. Appendix 29 is one example of my son's ideas.

AGE

Although age can hardly be classed as a "trait," it is rather interesting to note that innovative persons continue to exhibit creativity well into their 60s and 70s. Press* has referred to "the many important contributions made by scientists over the age of 55" and indicated that he "would be the last to tamper with this reservoir of talent." Similarly, Tsang† has pointed out that the innovative older scientists and engineers are often unsung heroes, acting as "big brothers" by passing on their ideas and analyses to their younger colleagues, citing Einstein's assistance to DeBroglie and Bose, and Rutherford's to Bohr. I can attest to Sir C. V. Raman's help to me, as a postdoctoral student, in the preparation of the paper chosen for inclusion in *Musical Acoustics*.

* F. Press, "Age and tenure," *Science (Letters)*, p. 219 (October 17, 1975).
† T. Tsang, "Creativity vs. age," *Physics today (Letters)*, p. 9 (August 1974).

The ages at which the brain *commences* its strong activity may also be worthy of examination. Thus it would appear that the earlier a child acquires the habit of activating his brain (the "cortical arousal" discussed in Chapter 1), the more adept he will be in maintaining that habit in later years (Appendix 30 and 31).

NEEDS

We close this chapter with some quotations from a report prepared by a committee of four Bell Laboratories engineers appointed by the then head of the Bell Laboratories Research Department, Mr. Ralph Bown. As noted in the opening sentence, the "Committee was appointed by Mr. Bown to study ways to stimulate fundamental invention within the Research Department." Its members were C. C. Cutler (who, with A. P. King and myself, prepared the Paris paper listed in Appendix Item 1), B. M. Oliver (mentioned in connection with pulse compression in the next chapter), W. A. Tyrrell (inventor of the hybrid or "magic" tee described in the next chapter), and myself (who was selected to be chairman). We list below some of the major recommendations:

RECOMMENDATIONS: 1. ENCOURAGEMENT OF INDIVIDUAL EFFORT

Although a well-coordinated group of scientists and engineers is essential to the successful development and application of new ideas, it is the individual who makes the fundamental discoveries and inventions which later form the basis for group activity. Even casual examination into the history of great inventions shows that highly creative thought focused upon a specific object is normally confined to the individual. If this be granted, it is clear that a stimulating research atmosphere is fostered only by keeping to a minimum those restraints and restrictions which tend to dull the edge of individual initiative. Beyond this, it is also important to provide a positive encouragement of the more talented staff members to remain alert for unexpected phenomena and new concepts which may emerge from their work and to pursue their new ideas to a definite conclusion.

1. Relax project system. In order to keep the work of the Research Department focused upon communications research, it has been customary to maintain a large portion of the effort in the form of projects. The project system provides the research man with a framework of subject matter and of problems in which he may exercise his originality. There is, therefore, genuine utility to be gained from projects in fostering new inventions and ideas. The progress of an invention beyond the original conception, however, depends in part upon the freedom accorded to the engineer for pursuing his ideas. If he is kept close to a project with certain rates of progress or deadlines to be met, it is improbable that his incidental ideas will receive further attention. It is suggested that the project system be sufficiently relaxed within the Research Department so that there is ample leeway for individual work in unexpected directions on the part of any engineer. This may necessitate some reduction in the total number of research projects so that each job will be adequately staffed.

2. *Instruct supervisors to stimulate invention.* The most outstanding discoveries and inventions cannot be classified as the end products of one or more stereotyped routines. On the contrary, each important original contribution has its own peculiar case history. From a comparative study of inventions, however, it is possible to formulate general conditions which often seem to form a background for important advances in scientific work. First, it appears advisable to have some framework of activity, some definite work to be done. In other words, it is very difficult to fabricate new ideas from thin air; some foreground of experimental facts must be present, preferably obtained by personal experimentation. Second, it is almost axiomatic that the research individual must display an alertness and a lively curiosity with regard to phenomena encountered in his work, especially irregular effects which cannot be explained by current knowledge and theories. Refusal to dismiss such side issues has often led to great things. Third, the individual should have the urge and drive to pursue new ideas and effects to the point where their merits are clarified with a sufficient degree of probability. To encourage the correct behavior on the part of the research worker, it is suggested that the responsibility be placed upon the supervisor for calling these matters to the attention of his engineers. He should see that each staff member understands clearly his responsibility to be alert for new ideas and concepts, observant of unusual experimental effects, and imbued with the desire to follow through to a probable conclusion. The supervisor should also endeavor to serve in this regard as a model for his staff by displaying impartiality in his own approach to new ideas, by encouraging the pursuit of new phenomena, and by generally making an obvious attempt to create an atmosphere conducive to research.

3. *Increase personal incentives.* The skilled scientific worker generally is a person with great deal of curiosity about the material world; if he did not have this inquisitive spirit, he would not have become and remained a scientist. Nevertheless, the average staff member in the Research Department is by no means indifferent to personal incentives, and he is more likely to put forth better effort if the possibility of personal gain or recognition is present. It is recommended that personal incentives toward creative work be increased in the two ways given below.

It is suggested that the promotion policy be such as to permit outstanding research men to receive the salary and prestige of supervisory rank without its executive responsibilities. This is to say, the staff member can look forward to advancement by virtue of good work and, when so promoted, he can continue his technical work without substantial supervision.

Quite apart from the incentive aspect, the present policy of rewarding high-quality technical achievement by the imposition of executive functions is directly responsible for removing some of the best people from technical research. It does not follow that a good research man will necessarily become a good administrative supervisor. It does follow that a successful research man who becomes an unsuccessful supervisor is misplaced so seriously as to depress the general efficiency of the organization, with the additional loss of his former technical ability.

It is suggested that more honor and distinction, both within and outside the Laboratories, come to those whose work is outstanding.

4. Periodically review personnel. Whenever it appears that a certain individual is not contributing his share toward inventive research, he should be given a different assignment or transferred to some other group. If it eventually turns out that he is not capable of a high degree of originality, he should be definitely assigned to the less creative phases of the work. Conversely, staff members throughout the Laboratories who show especial aptitude for invention should be given important research assignments in the Research Department whenever possible. It is suggested that simple routines be evolved to accomplish these ends.

5. Temporary transfers. More opportunities should be afforded to the technical staff to increase their capabilities and to broaden their funds of experimental techniques by means of temporary assignments to other groups within the Research Department as well as to related groups in the Apparatus and Systems Development Departments. While such transfers are not unknown at present, they have not been exploited sufficiently. It is suggested that supervisors be advised to be watchful for such temporary moves as may appear beneficial.

6. Provide sabbatical leaves with pay. Most of the technical staff do not have any effective chance to take leaves of absence to gain new perspectives and to receive specialized instruction from recognized authorities outside the Company. While it is formally possible to take a leave of absence without pay, this is not encouraged, and, in any case, the average engineer can seldom afford it financially. It is suggested that a system be set up similar to that of sabbatical leaves in use by the larger universities, whereby each year certain members of the research staff would be granted leaves of absence, with partial or full pay, to pursue selected studies at an appropriate educational institution in this country or abroad. In order to make the proposed system workable, it might be necessary to select only those who provide the most creditable plans for spending their leave and who are considered most likely to benefit thereby. The leaves might vary from three to twelve months, with each person potentially eligible for a leave every seven years.

The advantages to be gained from such a system are twofold: the individual will profit by the exposure to a new educational environment, while the group to which he returns will also benefit from his experiences and acquisition of fresh viewpoints.

7. *Improve coordination.* It is important that activities within the Research Department be coordinated to the point where the individual engineer is well informed on matters outside his own group. Such coordination should comprise the free exchange of information about projects and problems. It is believed that improvements can be made by supplementing the personal factor with the introduction of certain routines which operate automatically to insure continuity in coordination.

8. *Have more group meetings of the colloquium type.* Such meetings are a useful medium for the exchange and coordination of research results. They will be interesting in a casual way no matter how large the group may be, but they will usually be genuinely stimulating only when the group is small. It is preferable that the presentation by speakers be kept on an informal level, for, the more formalized the presentation, the less likely are problems and unusual results emerge as a stimulus to the audience. In a large colliquium, there is an unfortunate tendency to prepare talks in such finished form that they might equally well be delivered before the IRE or a similar outside society.

The Deal-Holmdel Colloquium may be cited as of an ideal size, while the West Street Colloquium is certainly too large. It is suggested that larger groups be broken up into several smaller ones, with the cleavages taking place along lines of interest or departmental organization.

As a final note, we include here the full statement of Alexander Graham Bell about diving into the woods, as it describes well some traits:

Don't keep forever on the public road, going only where others have gone. Leave the beaten track occasionally and dive into the woods. You will be certain to find something you have never seen before. Of course, it will be a little thing, but do not ignore it. Follow it up, explore all around it; one discovery will lead to another, and before you know it you will have something worth thinking about to occupy your mind. All really big discoveries are the results of thought.

With this subject now behind us, let us examine innovative developments in three more important fields of engineering, waveguides, lenses, and satellites.

6

WAVEGUIDES

EARLY RADIO TRANSMISSION

Just prior to World War II, the technology of radio had progressed to the point where very *short* radio waves could be generated. The first use of radio, in radio broadcasting, involved waves that were extremely long. We have all seen water waves generated when a pebble is dropped in a still pond. For such waves, the distance from one wave crest to the next (the wavelength) is perhaps a few inches. When we receive waves from radio broadcasting stations, those at the long-wave end of our receiver dial have wave crests separated by 1500 meters (almost a mile). As the technology for radio tubes developed, shorter and shorter radio waves could be produced. The forest of TV antennas that grew rapidly on the roofs of houses after World War II had *half-wave* dipoles (cross-members) which were only about 2 yards across, indicating that the wavelengths of TV radio waves were much smaller in length than those of the earlier broadcasting radio waves.

MICROWAVE WAVEGUIDES

By the late 1930s, radio waves could be generated having wavelengths only a few inches in length, and with this capability a new form of *transmitting* these waves become practical. The transmission structures for these very short waves were given the name "waveguides." These guides, which evolved mainly from two research groups, one headed by G. C. Southworth at the Bell Telephone Laboratories and the other by W. L. Barrow of the Massachusetts Institute of Technology. These waveguides were hollow conducting tubes, and their ability to conduct electrical (radio) energy seemed to upset the then popular ideas about the transport of electricity.

In the usual concept of electric "current," the energy is said to "flow" when a battery is hooked up to a flashlight bulb or some other device. The two wires employed in delivering energy to the bulb were called the "direct" and "return" leads. Current could flow to the device over the direct lead and back to the battery over the return lead.

For alternating current, we learned that things are only a little different. A toy electric train transformer circuit, which uses alternating electric current, cannot be said to have a return lead, since both leads, because of the "back-and-forth" nature of the flow, act first as current "suppliers" and then as current "returners." Nevertheless, such circuits still have two leads. So does a telephone circuit carrying currents which alternate back and forth at several thousands of times per second. Even the extremely high-frequency transmission circuit, the coaxial cable, which carries currents with frequencies of cycles per second, possesses, in addition to one central conductor, an outer cylindrical conducting shell that can be looked upon as constituting the second "wire" of the circuit.

Accordingly, when the waveguide developers announced the transporting of electrical energy over a single conductor, a hollow metal tube, obvious questions were asked. Where is the return circuit? If the current travels out in the tubular conductor, how does it get back? Of course, when waveguides were described to physicists experienced in optical matters, no such consternation resulted. They recognized that radio waves and light waves, both electromagnetic, would behave alike, and the idea of passing light waves down a tube seemed not at all radical. Thus a waveguide can be considered not just a conductor of electricity but rather as a confining structure for propagating electromagnetic waves. The wavelengths of the radio waves involved are so short that they behave like light waves. Figure 67 shows microwaves issuing from a rectangular waveguide.

THE WAVEGUIDE AS A TRANSMISSION MEDIUM

When radio waves are propagating within the enclosed conducting structure of a waveguide, the usual energy distribution causes their

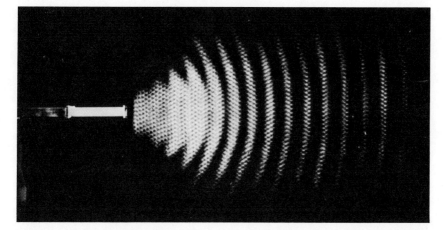

Fig. 67. Single-wavelength microwaves issuing from a waveguide (the procedure for visually portraying patterns such as this is described in the author's books Radar, Sonar and Holography (Academic Press 1973, pp. 110ff) and Seeing Sound (Wiley, 1971, pp. 7–32).

intensity to be strong at the center and weak at the walls. This simplest energy configuration is called the "dominant" configuration or, more usually, the "dominant transmission mode." It is shown in Fig. 68 for a rectangular waveguide. The metal side walls "short-circuit" the electric field, and it falls to zero there. Waveguide transmission systems now in common use can pass radio frequencies ranging from about 1000 million to over 100,000 million cycles per second (1 to over 100 gigahertz).

Waveguides permit the transmission of radio waves over moderate

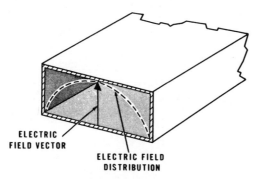

ELECTRIC
FIELD VECTOR

ELECTRIC FIELD
DISTRIBUTION

Fig. 68. The electric field in a rectangular waveguide is strong in the center and zero at the edges.

distances with relatively low losses. Also, the conducting tube acts as a very effective shield and prevents interaction between very-high-energy signals propagating in one waveguide and low-amplitude signals in another waveguide nearby.

PHASE AND GROUP VELOCITIES

A strange thing happens to electromagnetic waves confined within a waveguide; transport of the energy slows down but the wave velocity increases. When electromagnetic waves travel unhindered in free space, the velocity of propagation of energy is the same as the speed with which the crests and troughs of the waves move. But in waveguides these two velocities must be differentiated, for they are not the same. We call the velocity of propagation of the energy the "group velocity," and the speed with which the crests and troughs of the waves themselves move the "phase velocity."

In later discussions of this wave property within a waveguide, in connection with waveguide lenses, we shall be particularly interested in the wave or phase velocity. Since wavelength is directly proportional to velocity, a higher wave velocity within a waveguide must mean a longer wavelength within the guide. The free-space wavelength is usually designated as λ_0 and the wavelength within the guide as λ_G. The "stretching" of the wavelength of a radio wave entering a waveguide is illustrated in Fig. 69.

Figure 70 is a ripple tank photo of water waves, simulating the case of microwaves propagating in a waveguide. The channel causes the water waves to have a phase velocity that is dependent on wavelength, just as the phase velocity of microwaves in a waveguide is dependent

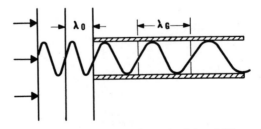

Fig. 69. Free-space radio waves arriving from the left exhibit an increase in wavelength upon entering a metallic waveguide.

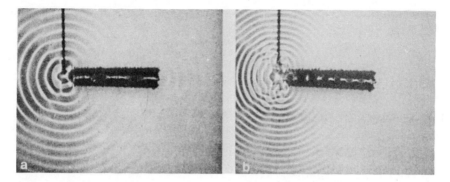

Fig. 70. Photos of a ripple tank showing (a) the increased wavelength (increased wave velocity) of waves entering a waveguide and (b) the smaller increase occurring when the wavelength of the entering waves is shorter. Courtesy Allen H. Schooley, U.S. Naval Research Laboratory.

on wavelength. In Fig. 70(a), the circular pattern shows fairly long-wavelength waves radiating from a point source and then entering the waveguide structure. For this long-wavelength case, the "guide wavelength" at the right is seen to be much longer than the wavelengths of the circular pattern. In Fig. 70(b), the wavelength has been made shorter (as is evident from the circular pattern), and the change in the wavelength within the guide relative to the circular-pattern wavelength is seen to be much smaller than the change shown in Fig. 70(b).

LONG-DISTANCE TRANSMISSION

The waveguide experienced extensive use in radar, but it also provided a new communications possibility for the transmission of high-information-content signals, presently accomplished largely by coaxial cable and radio relay circuits. (We discussed in Chapter 4 the contemplated use of light fibers for a comparable application.) As early as 1936, G. C. Southworth introduced his discussion of "electric waveguides" (in the Bell Laboratories Record) with the following:

> Researchers in the Bell Telephone Laboratories have disclosed a new form of transmission for high frequencies. It is unlike radio because the waves are not broadcast through space but follow a physical guide comparable to a wire. No return path, however, is required of the kind that is commonly assumed in the usual case of transmission.

However, as has been noted, "four decades of increasing demand and increased sophistication in electronic equipment had to pass before this new medium could come into its own in a long distance transmission medium."*

MILLIMETER WAVELENGTHS

The improvements in electronic equipment referred to included the development of extremely-high-frequency microwave generators and amplifiers, for it is in that frequency range (millimeter waves as contrasted with the centimeter radio waves of radar and radio relay circuits) where waveguide transmission techniques shine. The reason is that the very short millimeter waves cannot be used in radio relay or earth-to-satellite applications because these waves are strongly attenuated by rain and even fog. Within the waveguide, this rain problem obviously does not exist, and the very short wavelengths of millimeter waves correspond to very high frequencies, thereby permitting the use of much wider bandwidths and hence more communication channels. Thus in one system under test by the Bell System, called the "WT4," the millimeter waveguide can handle 230,000 simultaneous phone calls.

THE CIRCULAR ELECTRIC MODE

It was recognized quite early that the standard radar waveguide mode (the "dominant" mode of Fig. 68) would not be usable in long-distance circuits because of its rather high attentuation per mile, resulting in the requirement of far too many amplifier stations (as compared, for example, to coaxial cable).

Here again, however, a multidisciplinary innovator had come to the rescue, *theoretically,* and quite early. The brilliant Bell Laboratories mathematician Sergei Schelkunoff had become knowledgeable enough in another field, that of microwave concepts and requirements, to be able to apply his math aptitude to the analysis of waveguide attenuation. In that early "exercise," Schelkunoff came up with a

* William D. Warters, "Millimeter waveguide scores high in field test," *Bell Laboratories Record,* pp. 400–408 (November, 1975).

(theoretical) prediction that truly did provide (much later) the only effective solution to the waveguide attenuation problem. He showed mathematically that a waveguide mode in a circular (cylindrical) waveguide would exhibit extremely *low* attenuation, in contrast to the dominant mode of Figure 68. Schelkunoff made his analysis in the 1930s,* but for several years after I joined the Research Staff of the Bell Telephone Laboratories in 1942 (my first assignment was at the Holmdel Microwave Radio Research Laboratory) convincing proof of Schelkunoff's prediction was still not available. World War II was then demanding maximum consideration of radar applications by microwave specialists, and it was not until after the end of the war that a completely accurate test of Schelkunoff's prediction was made. As any mathematician would have predicted, Schelkunoff's theory was fully corroborated, and today field tests are proving the value of his interdisciplinary capability.†

RECENT FIELD TESTS

Naturally the step from theory to practice was not an easy one, because, to obtain the theoretical loss of only 1 decibel or so per mile (stated more simply, this means that the input power drops to one-half its value after it has traveled 3 miles), many precautions must be taken. The circular waveguide diameter needed to achieve this low loss is so large that it permits literally hundreds of other (much higher-loss) modes to propagate within the guide so that the waveguide design must be such as to minimize the generation (by conversion from the low-loss mode) of these unwanted modes. In 1952, I was issued a patent on the suppression of such modes (Appendix 32). More recently, Bell innovators have developed waveguide designs to *fully* meet this requirement, one such design being a waveguide having a thin, low-loss dielectric lining and another being a waveguide having a thin *high-loss* dielectric and also having a thin copper wire wound helically on the surface of the dielectric. In the installation over which tests were recently conducted (an 8¾-mile route), over 98 percent of the waveguide was dielectric lined (low-loss), the remainder being short sections of the (high-loss

* "Forty years of waveguides: A glimpse of history," *Bell Laboratories Record* (March 1970).
† Warters, *op. cit.*

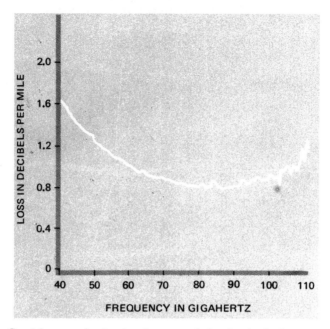

Fig. 71. Crucial test results showing the extremely low loss in the frequency range from 40 to 100 gigahertz over the 8¾-mile test route. Courtesy Bell Laboratories.

dielectric) spiral wire type (which filtered out any unwanted modes that may have been generated). Once installed along the route, the signal loss over the full 40- to 110-gigahertz band could be experimentally measured (Fig. 71), and it was found to be lower than that originally predicted (for that design) in 1972—low enough to permit repeater stations to be spaced up to 25 miles apart.*

WAVEGUIDES AND COAXIAL CABLES

The 230,000 (phone) circuit capacity of the WT4 waveguide is quite large by today's Bell standards. Thus the first (L1) coaxial cable carried 480 circuits per coaxial pair, and the first L5 cable, which entered service in 1974, carried 10,800 circuits per pair. And experiments at Bell are now under way to increase (through a special form of phase modulation) the waveguide's capacity to 460,000 circuits.† Quite

* Warters, *op. cit.*
† Warters, *op. cit.*

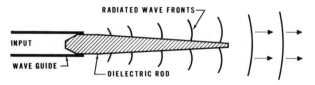

Fig. 72.   A tapered dielectric rod, which not only guides microwaves (from a metal wave-guide) but also radiates them gradually, constitutes a useful end-fire radiator.

a tribute to electrical engineers Barrow and Southworth, and especially to mathematician Schelkunoff.

DIELECTRIC WAVEGUIDES

Quite early in the microwave radar picture, the use of a *dielectric* antenna was found to be advantageous. Figure 72 shows how a tapered dielectric rod can radiate microwave energy in the forward direction so as to become an "end-fire" antenna (radiator and receiver). Such dielectric rods (Fig. 73) were employed in an important early World War II radar (Fig. 74). If, instead of reducing the cross-section of the dielectric rod, as in Fig. 72, the rod is continued at full diameter, it can act as a waveguide, with its "guiding" property being a consequence of

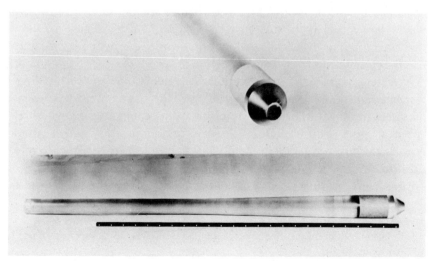

Fig. 73.   A dielectric rod used in an early World War II radar.

Fig. 74.   The radar which used 42 of the dielectric waveguide rods in Fig. 73 as a steerable array radar.

the lower wave velocity in the dielectric.* In the following chapter, we shall discuss the use of *artifical* microwave dielectrics, for use in microwave lenses.

METALLIC DIELECTRIC WAVEGUIDES

One of the varieties of artifical dielectric lenses discussed in the next chapter utilizes closely spaced arrays·of metallic disks. If these disks are mounted on a rod, as shown in Fig. 75, they act on microwaves very much as the dielectric rod of Fig. 72 does. Figure 76 shows that a *long* rod of disks can conduct microwaves over extended distances much as a hollow waveguide does.

A few years ago, a dielectric waveguide called a "surface wave transmission line" was investigated by the Environmental Science Services Administration as a way of providing communications to high-speed trains.† Full radiation of such communication signals would require permission to employ frequency bands already assigned to other uses, so the waveguide signal, which remains strong only within 20 or 30 feet from the guide, does not violate (FCC) requirements, yet can be

* W. E. Kock, *Sound Waves and Light Waves* (New York: Doubleday, 1965).
† J. B. Scott, "Surface waves find a job chasing high-speed trains," *Microwaves*, pp. 7–12 (September 1970).

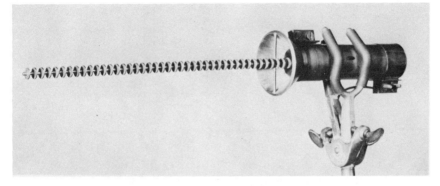

Fig. 75. A long row of conducting disks constitutes an "artificial dielectric" rod waveguide radiator similar in action to the true dielectric rod in Fig. 73.

received by trains traveling on tracks along which the waveguide is mounted. In the prototype version, the guide is a 1-inch-diameter copper tube coated with a 0.32-inch polyethylene covering which guides and *confines* the wave energy. It is positioned 6 feet above the ground (Fig. 77).

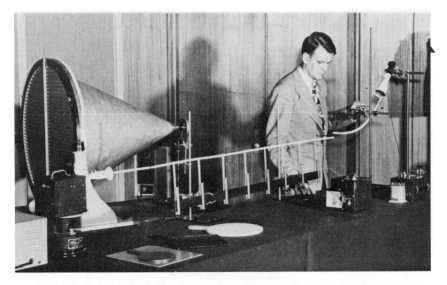

Fig. 76. The disk-on-rod artificial (metal) dielectric can be extended to quite long lengths. The "white wire" in this photo acts as a waveguide for microwaves.

Fig. 77. This prototype of a communications system to high-speed trains employs a dielectric "surface transmission" waveguide. Courtesy U.S. Department of Transportation.

ELECTRON BUNCHING

During the late 1930s, extensive research efforts were directed toward ways for generating the very short-wavelength radio waves called "microwaves." One way which proved very successful, and which was developed in several laboratories in the United States and Europe, was called "electron bunching." For the many scientists and engineers who were not specialists in electron tube technology, a paper by two German scientists, E. Borgniss and E. Ledinegg, provided an easily understood explanation of how such a process could generate microwaves. Their paper, published in 1940,* described the bunching of electrons in terms of a change in the velocity of the electrons.

BUNCHERS AND CATCHERS

Figure 78 shows the structure of the electron tube used to generate the short-wavelength radio waves by electron bunching. Electrons origi-

* E. Borgniss and E. Ledinegg, *Z. Tech. Phys., 21,* 256 (1940).

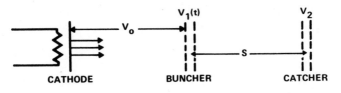

Fig. 78. By modulating at the "buncher," the velocity of electrons can be made to arrive at the "catcher" simultaneously.

nating at the cathode at the left of the figure move to the right with velocity V_0 and pass first through two structures ("grids") which constitute the "buncher." The grids are able to impart different speeds to the electrons as they leave the second grid. This speed control or "velocity modulation" is accomplished by altering the (axial) voltage, $V_1(t)$, which exists across the two grids. Accordingly, if this voltage is made to *vary* in such a way as to cause those electrons which leave the buncher at a later time to acquire a higher velocity, these later electrons can be made to overtake electrons which left at an earlier time (with a lower velocity). Figure 79, plotting distance vertically and time horizontally, shows how the velocity difference between two groups of electrons, one leaving at time t and the other at a slightly later time $t + \Delta t$, causes the two groups to arrive simultaneously (at time t_1) at the same axially distant point S_1, which, let us assume, corresponds to the distance S in Fig. 78. In that figure, there is, at a distance S from the buncher, a "catcher." If a series of bunched (concentrated) groups of electrons can be made to pass *its* two grid structures periodically, a

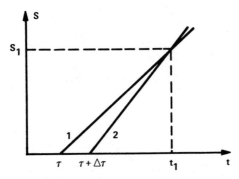

Fig. 79. Electrons leaving at an earlier point in time (τ) can be overtaken by higher-velocity ones leaving at a later time ($\tau + \Delta\tau$).

device connected to them, called a "toroidal resonator," will generate
the desired microwave radio waves.

IDEAL BUNCHING

Figure 80 portrays, in the bottom graph, how the voltage (V) must
vary ideally with time (t) in order to achieve the most effective bunching. (Figures 78, 79, and 80 are taken from a paper I wrote late in
1941.) The middle graph shows how repeated sets of four electron
groups (the groups correspond to the sets of four tilted lines which are
periodically repeated) are thereby made to arrive as a single, repeated
(periodic) groups at the catcher distance (S). In the top graph, this
bunching is shown as a variation of current (i) with time, indicating the
compression (with time) that *theoretically* can be made to occur.
Because of the very high frequencies (billions of cycles per second)
involved in the generation of the short-wavelength microwaves, it is
almost impossible to produce for them the ideal voltage variation across
the buncher grids such as that sketched in the bottom graph of Fig. 80.
The usual voltage variation is sinusoidal (sinuous), resulting in lesser
amounts of compression (bunching).

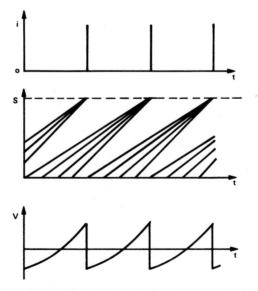

*Fig. 80.   Velocity variations (bottom) imparted to electrons (middle) cause them to
arrive together (top).*

KLYSTRONS

The velocity modulation principle just described became widely used in microwave tubes called "klystrons." Some used the direct, two-structure buncher and catcher process of Fig. 78, while others used a rather clever variation in which the electrons, after passing the buncher, were caused to be reflected (their direction of motion reversed) so as to *again* pass the original buncher. That form of electron tube was referred to as a "reflex" klystron, and it became quite important in microwave technology by serving both as an "oscillator" and a microwave *generator*. These reflex klystrons in which the electrons were made to reverse their direction enabled the electrons on their return to excite, not a catcher as in Fig. 78, but the original buncher structure, causing it to become more effective in bunching *new* electrons arriving from the cathode. These new electrons were similarly reflected, causing the process to build up in time to such an extent that an oscillatory condition resulted.

The effect can be likened to the oscillations of air pressure in an organ pipe; the pressure of the reflected pulse of air from the top end of the organ pipe affects the air being blown into the input end, and the pipe "oscillates," that is, it "speaks" or sounds a musical tone. The frequency, or pitch, of the note sounded by the organ pipe can be altered by altering the point at which the pulse of air is reflected, that is, by altering the *length* of the pipe and thus shortening the travel time of the pulse. Accordingly, the longer the organ pipe, the lower its frequency. In a similar way, the frequency of the oscillating radio signal generated by the reflex klystron could also be changed by means of a length change, by altering the effective position of the reflecting plate which reverses the direction of the electrons. This was done electronically (with no actual movement of the reflector plate) simply by altering the voltage applied to the plate. This feature and others made the reflex klystron an extremely useful microwave oscillator.

VELOCITY MODULATION OF WAVES

For a number of years, the concept of velocity modulation, or bunching, remained confined to electrons. But shortly after World War II some off-the-track thinking began to emerge. We saw in Fig. 70 that the velocity of microwaves in waveguides was dependent on their wavelength (their frequency). We noted at that point that for microwave radio waves traveling inside a waveguide the *energy* velocity (or, as it is more usually called, the group velocity) is *lower* than its value in free space (that is, outside the guide), whereas the *wave* velocity (the speed of travel of the wave crests) is higher. We here consider how *wave energy* can be bunched through the use of the change in the velocity of waves in a waveguide when the frequency (or wavelength) of the waves is changed.

In January 1947, I entered in my Bell Laboratories notebook (Appendix 33) a discussion, not of *electron* bunching, but of *wave* bunching or "photon bunching," as I referred to it. This entry described "The idea of using a waveguide as a means of changing the velocity of microwaves by varying their frequency and thus [producing] a 'velocity modulation' effect, whereby the individual energy packets fed into a waveguide at different times can be made to catch up with one another." Since it is the energy which is to be bunched or concentrated, this entry referred to the energy (or group) velocity. Further along, the notebook entry observed that the higher-frequency wave packet can enter the waveguide input "at later time but due to higher group velocity it arrives at catcher point coincident with f_0 [an earlier, lower-frequency wave packet]."

RADAR BREAKDOWN PROBLEMS

It was about this time (1947) that radar transmitting tubes were reaching such high output powers that for the short pulses then being used to acquire a high resolution of targets overloading (breakdown) of transmitter portions of radar systems occurred. In order to put *more* power into the transmitted radar pulses, the only alternative was to increase the pulse *length*. Such an increase in the pulse length of a sonar "ping" is illustrated in Fig. 81. It is evident from the figure that this increased length prevents the radar (or sonar) system from "resolving" two targets located close together, targets spaced .apart, say, at a distance corresponding to twice the length of the shorter pulse. This is because the longer pulse system would have its received pulses overlapping and the problem of identifying the location of the two targets would be extremely difficult. Obviously if there were some way of "coding" the long outgoing pulse of Fig. 81 so that, upon reception of it as a target echo, it could be "decoded" and thereby compressed in length (so

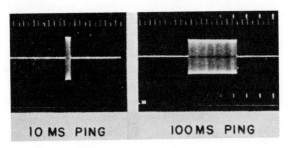

Fig. 81.   Sonar pulses ("pings") of different lengths.

as to become similar to the short pulse of Fig. 81), it would provide a simultaneous solution to the breakdown problem and to the problem of resolving closely spaced echoes.

WAVE CODING

Now the wave energy velocity modulation discussed in my 1947 notebook entry (Appendix Item 33) provided just such a coding and decoding process. The outgoing (long) pulse could be made ("coded") to consist of a signal whose frequency varied with time during the duration of the pulse. Upon reception, after reflection from a target, the frequency-varying (reflected) pulse could be introduced into a waveguide (or a comparable filter in which the energy velocity was dependent on the frequency as in a waveguide), thus causing the later portions of the pulse (having higher frequencies) to catch up to the earlier (lower-frequency) portion. This waveguide "decoding" would thus cause a shortening (length compression) of all *received* pulses, permitting the reflections from two closely spaced targets to be resolved. It would be comparable to the situation in which the transmitted pulses would be *long* (high-power) pulses (as at the right of Fig. 81) and the "decoded" compressed pulses would be short (as at the left of Fig. 81).

CORRUGATED SOUND-WAVE WAVEGUIDES

As a convenient way for experimentally verifying the pulse compression concept, my colleague F. K. Harvey caused pulses of the proper-frequency sound waves to propagate in a cylindrical tube having a corrugated wall. The corrugations were approximately one-quarter wavelength deep for the sound waves employed so that the tube acted, for sound waves, just as the waveguide of Fig. 68 acted for microwaves. A zero value of sound pressure was caused to exist at the "wall" by the quarter-wave corrugations, just as the side wall of the waveguide of Fig. 68 "short-circuited" the electric field (causing it to be zero) in the case of the conducting (copper) waveguide. The velocity of propagation of the sound waves was thus dependent on frequency as in the waveguide, so that a pulse of higher-frequency sound waves could "catch up" to a pulse (introduced earlier) of lower-frequency sound waves.

PULSE COMPRESSION TESTS

Floyd Harvey and I chose an equivalent test: a single acoustic pulse *comprising two frequencies* was introduced into the corrugated-wall tube and pickup microphones were placed at two points along the tube (Appendix 34). Figure 82 shows how the two originally super-imposed pulses become separated as they travel down the tube. This experiment in *reverse* constitutes the pulse compression process and permits a long transmitted pulse which has a rising frequency (not unlike the rising "chirp" of certain bird calls) to be coalesced into a short pulse when received as an echo.

CHIRP

At about this time, another of my colleagues at the Bell Labora-tories, Barney Oliver, issued a memorandum whose title was reminis-cent of the line "Not with a bang but a whimper" from T. S. Eliot's 1925 poem The Hollow Men. Just as the outgoing sonar pulse (Fig. 81) was called a "ping," the outgoing (extremely-high-power) radar pulse was generally referred to as the "bang." Oliver accordingly entitled his memorandum "Not with a Bang But a Chirp." But it remained for Bell Laboratories scientist Sid Darlington to fully work out the chirp pulse compression details,* and it is he who is given the most credit for the development of the technique. Figure 83 shows how effective the process can be. Recently, using the new optical discovery called "holography,"

* J. R. Klauder, A. C. Price, S. Darlington, and W. J. Albersheim, "The theory and design of chirp radar," *Bell System Tech. J., 39,* p. 745 (1960).

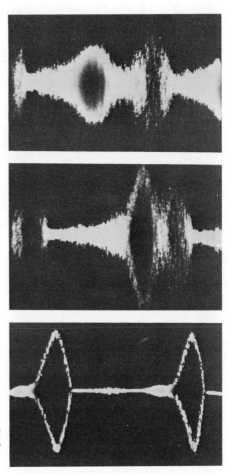

Fig. 82. Two superimposed pulses, each having a different frequency, are shown at A as they enter an acoustic waveguide in which the sound velocity depends on frequency. This dependence of (group) velocity on frequency causes the two pulses to become separated in space (and time) as they move along the tube (B and C).

another innovator, Fordyce Tuttle, suggested a holographic technique for accomplishing pulse compression. So, with a little diving into the woods, World War II electron bunching became a boon to the more recent, more sophisticated technology of radar and sonar.

NATURAL WAVEGUIDES

During the early days of World War II, radar operators began to notice occasional instances during which the useful range of their radars (against aircraft targets) increased very significantly. Planes were often used as targets for determining the maximum range capabilities of a

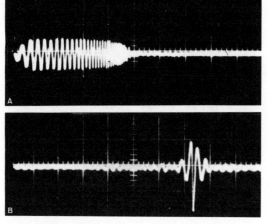

Fig. 83. When the long acoustic pulse (A) in which the frequency rises with time (as in the "chirp" of a bird call) is passed through a device in which the velocity depends on frequency (as in Fig. 82), its length is appreciably compressed (B).

given radar system, and, during these unusually high-range situations, the target plane, upon exceeding the design range, could still be clearly observed, on what was called the "second trip around" of the radar's range display. Calculations showed that such phenomena could not be explained if the radar radio waves were assumed to be propagating in the usual manner (that is, through a uniform atmosphere) because the radar's transmitted and received powers were not large enough to account for this excessive range capability.

Now it was known that radio waves do suffer a slight reduction in their velocity when they pass through air having a high moisture content (a high humidity). This slowing-down effect is similar to the case of radio waves or light waves passing through glass (as in a lens or prism), where their velocity is sizably reduced. Someone therefore began to envision a situation where the air's moisture content was low at the earth's surface, high at higher altitudes, and low again at still higher altitudes. This particular arrangement or structure of the atmosphere would provide high-velocity regions above and below the middle (high-humidity, low-velocity) region (the "axis" region). Such a structure, in which the propagation velocity is low at the axis but increased gradually with the distance out from and on both sides of the axis, constitutes a guiding means. It continually refocuses, back toward the axis, energy which may be proceeding away from the axis. This effect is sketched in Fig. 84; at (a), the "tapered" velocity structure behaves like a series of lenses, as sketched in (b).

Accordingly, when the moisture vapor content in the atmosphere

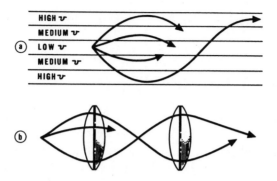

Fig. 84. The proper variation of velocity within a region can create a waveguiding effect. The region then behaves like a series of lenses.

above the earth's surface varies with height, as described above, a pronounced guiding effect results for radio waves, particularly for those having the shorter, radar wavelengths. When this effect is present, unusually long-range transmission occurs. The upper portion of the sketch in Fig. 85 suggests how the performance of the radar shown is improved when a strong meteorological waveguide "trapping" effect exists; aircraft can then often be detected at ranges so great that, for the radar location, these planes are "over the horizon." Under normal conditions, aircraft or other targets beyond the horizon cannot be "seen" because the microwaves used in radar, like light waves, travel in

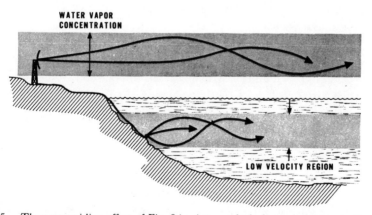

Fig. 85. The waveguiding effect of Fig. 84 exists regularly for sound waves in the deep ocean and occasionally for radio waves in the atmosphere.

straight lines, and the earth's curvature produces a strong shadowing effect.

UNDERWATER WAVEGUIDES

It was at this point where serendipity entered the picture. During World War II, numerous scientists were working in the field of underwater sound and in its application to acoustic detection systems, called, analogously to the radar detection radar systems, "sonar systems." One of these experimenters, the U.S. scientist Maurice Ewing, learning of the natural (meteorological) radar waveguides, began to wonder if such a wave guide effect might exist for the sound waves in the ocean. Ewing knew that because of the increasing weight of the water above, the water *pressure* in the ocean constantly increases with depth. On the other hand, the water *temperature* is warmest at the surface (in most deep-ocean, temperate-climate areas) and it *decreases,* up to a point, with depth. Eventually, at a depth of about 4000 or 5000 feet, the temperature remains constant with increasing depth. Now this situation is exactly what is needed for the formation of a *permanent* acoustic waveguide. In the region near the surface the higher temperature causes the sound velocity to be higher, and below the point where the temperature becomes constant with increased depth the still increasing pressure also causes the velocity to be higher. In between these there is a lower-velocity region, and as in Fig. 84(a) and as shown in the lower part of Fig. 85, these regions in the ocean provide a waveguide effect for sound waves.

Accordingly, in the deep portions of the oceans the variations of the water temperature with depth, combined with the effect of gravity, create a sound "channel" which guides underwater sound signals to very great distances.

THE SOFAR CHANNEL

By virtue of the existence of a sound channel in the deep oceans, underwater sound signals have been transmitted and received over paths many hundreds of miles long. When the United States exploded a nuclear bomb in the deep water of the Pacific Ocean (Project Wigwam), *echoes* of the detonation (corresponding to reflections of the sound from the coasts of China and Japan) were "heard" off the coast of California. Meteorologists have also found this deep sound channel useful. Underwater listening devices can detect the noises made in the central, highly disturbed portion of a typhoon or hurricane hundreds of miles away. The course of the typhoon can be followed by noting the direction or bearing changes of the sound source. The deep sound duct in the ocean was named, by its discoverer, Maurice Ewing, the "SOFAR channel," from the words *SO*und *F*ixing *A*nd *R*anging. Appendix 35 is a photo of Ewing taken when he visited me at the Bell Laboratories. The picture was taken in the quiet, "free-space" anechoic room there.

SOFAR LOCALIZATION

The discovery of the SOFAR channel by Ewing was quickly followed by his recognition of a valuable use of it, that of locating downed aircraft personnel. This use involved the rather simple installation of three or more acoustic receivers (hydrophones), positioned deep enough to benefit from the 5000-foot-deep SOFAR channel sound propagation characteristics and connected by cable to shore. Aircraft pilots were provided with "SOFAR bombs," which they threw overboard from their liferafts. As the bomb sank, the weight of the water above it increased, and when it reached the SOFAR channel depth the water pressure became large enough to trigger a device which exploded it. This explosion created a large sound disturbance within the SOFAR channel, and the time of arrival of this explosive sound at the three receiving stations provided the stations with an excellent position fix on the spot where the bomb was dropped. (This follows from the fact that a sound travels outward as a circular disturbance, and its time of arrival at three different receiving locations defines the point where the explosive sound originated.) Shore stations could thus locate the position of the downed flyer and send out a rescue mission.

NONEXPLOSIVE ACOUSTIC USES

In recent years, the sound transmission properties of the SOFAR channel have been investigated at some length. Figure 86 shows an array of underwater loudspeakers being lowered into the Atlantic Ocean for conducting such experiments. These units can transmit (radiate) a wide band of sound frequencies quite efficiently; they were designed by U.S. scientist G. S. Bennett of the Bendix Research Laboratories. The unit in Fig. 86 was later installed permanently on the ocean bottom off the northern shore of the Island of Eleuthera, an out-island of the Bahamas. Figure 87 shows it, attached to a rigid base structure, being lowered into the sea, and Fig. 88 shows the cable to it being brought ashore. Open Atlantic Ocean lies between this location and the south shore of Bermuda, some 700 miles away, with the water depths in between exceeding the SOFAR channel depth over the entire distance. Bell Laboratories Scientist R. H. Nichols, using a deep water listening system connected by cable to a shore laboratory on Bermuda, conducted extensive tests, over long periods, on the sound transmission properties of this "acoustic link." Because the speed of sound in water is about a mile per second, a sound produced by the unit at Eleuthera takes about 700 seconds or almost 12 minutes to travel to Bermuda.

Fig. 86.   An array of underwater loudspeakers about to be lowered into the ocean.

Fig. 87. The array of Fig. 86 now configured for placement on the ocean bottom, in deep water.

Fig. 88. The connecting cable to the bottom-mounted array of Fig. 87 is brought to a shore station by a series of floats.

ACOUSTIC ATMOSPHERIC WAVEGUIDES

The variations in moisture and air density in our atmosphere also affect the propagation of sound waves (in air). As in the case of radio wave propagation discussed earlier, the sound velocity in the atmosphere also depends on the moisture content and density. At a rather high altitude there exists in the earth's atmosphere a relatively permanent "sound channel" where sound is continually refracted back toward the "sound axis," as were the radio waves of Fig. 84. Exceedingly loud sounds—heavy explosions, for example—can be "heard" at great distances because of this acoustic waveguide in the atmosphere. Large nuclear explosions set off above ground are detectable thousands of miles away; Figure 89 shows the acoustic record of a nuclear explosion as received 11,500 kilometers away from the blast. The word "heard" is put in quotation marks because the sound waves detected are only the very low-frequency sound waves. Because these cannot be heard by our ears, special low-frequency detection equipment (microphones) must be used.

WAVEGUIDE JUNCTIONS

We saw earlier that when radio had eventually developed to the point where very *short* waves could be generated, a new form of transmission structure became feasible, the hollow metallic tube waveguide. We also saw that waveguides exhibit the rather unusual property that the wave velocity of radio waves propagating within the guide is *higher* than the velocity of the same waves propagating in free space.

Because the more usual configuration of these wave "conductors" involved rectangular guides, several experimenters noticed that a con-

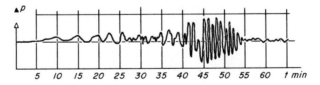

Fig. 89. The atmospheric "waveguide" can cause loud sounds to be "trapped" and thus heard many miles away. This is an acoustic record of the airborne signal of a nuclear explosion. After L. Brechovskich.

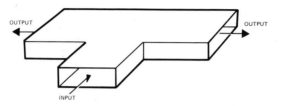

Fig. 90.   A waveguide "dividing" junction.

nection from one rectangular guide to two other such guides could be made in two fairly straightforward, fairly natural ways. One involved the connecting of the three guides in the *flat* dimension as shown in Fig. 90, and the other had the junction leg perpendicular to the flat guide dimension (Fig. 91).

Now the average person would look upon these two as being rather similar. Certainly if one caused, say, water, to be forced into the input leg of the unit of Fig. 90, it would have to exit equally at the two output points where arrows are shown, and similarly for the unit shown in Fig. 91.

TEE JUNCTIONS

But *some* waveguide experimenters knew that radio waves had certain special properties that could make a difference in these two types of junctions. (Both arrangements were given the name "T-junctions" or "tee junctions.") In the sketch of the waveguide in Fig. 68, there is a line referred to as the electric field vector, and it has an arrowhead pointing upward on its upper end. This is to indicate that the radio waves involved have a preferred orientation (a "polarization").

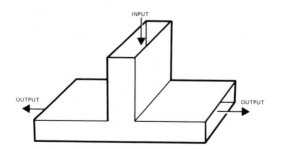

Fig. 91.   A second type of waveguide "dividing" junction.

Let us assume that the wave energy entering the input guide of Fig. 90 has the same arrow orientation (the same polarization) as that indicated in Fig. 68. It would then be reasonable to expect that this energy would exit at both outputs (shown by the horizontal arrows) with the same orientation, that is, with the arrow pointed upward in both branches. Experiment shows that this is indeed what does happen.

Now, however, let us examine what happens when the energy enters the input waveguide of Fig. 91. The arrow of Fig. 68 would now be *horizontal* at the input, and there would be some question as to what its orientation would be at the two exit ports (output points). If we assume that, at the input in Fig. 91, the arrow points to the left, one would expect that, as the arrow "walks down and around" toward the left-hand output, it would be pointing upward at the output point (as in Fig. 68). If we follow a similar "walking down and around" analysis, however, for the right-hand output, we find that when the arrow at the input points left, it would walk toward the right with the arrow pointing down. Experiment shows that this difference does indeed occur.

In technical terms, the tee of Fig. 90 has been referred to as an "H-plane tee" or "shunt-junction tee," and when such a dividing junction is properly "matched," energy fed into the input arm will divide equally and be in *equal phase* in the two output arms of the tee.

The tee in Fig. 91 is called an "E-plane tee" or "series junction tee," and when that junction is properly matched, energy fed into its input leg will divide equally in its two output arms, but, as suggested from our analysis above, the energy in those two output legs will be 180 degrees *out of phase* (the arrow of Fig. 68 will be pointing upward in one and downward in the other).

HYBRID JUNCTIONS

For some time this difference between these two junctions was widely known. Accordingly, it took a little serendipity, or off-the-beaten-track thinking, to arrive at the *next,* extremely important form of waveguide tee. It was apparently envisioned independently at the two important U.S. World War II microwave research establishments, which, as we mentioned earlier, were the Bell Laboratories and the Radiation Laboratories at the Massachusetts Institute of Technology. However, Bell Laboratories' scientist Dr. Warren Tyrrell obtained the basic patent on this newest waveguide junction, and this suggests that his discovery of it was the earlier.

Because it was recognized as being extremely important to waveguide technology at both laboratories, it became widely known and widely referred to at each of the two laboratories. Because of the somewhat conservative nature of Bell Laboratories, it was there referred to simply as a "hybrid waveguide junction" or simply as the "hybrid junction." The researchers at MIT on the other hand, were so intrigued with this very interesting device that it was there given the name "magic tee."

THE MAGIC WAVEGUIDE JUNCTION

Figure 92 shows the construction of this magic waveguide device. When properly matched, energy fed into input leg 1 divides (as in Fig. 90), equally and *in phase,* in outputs 3 and 4. In addition, however, through microwave magic (?) *no energy enters the vertical leg, number 2!* Similarly, an input into leg 2 divides equally (but with *opposite* phase) into legs 3 and 4, and again magically (?) *no* energy enters leg 1! Thus, for the matched conditions, *there is no coupling between legs 1 and 2.* Truly a magical situation.

This hybrid junction has since found innumerable extremely important applications in waveguide technology, in both the communications and radar fields. One important application occurred in a form of radar referred to as "monopulse radar." In that radar, the two legs 3 and 4 of Fig. 92 were extended and curved around in their same plane so that their exit points were side by side (that is, behind the device as

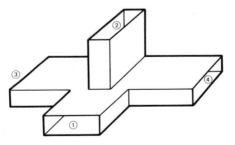

Fig. 92. A waveguide double-dividing junction, called a "hybrid junction" or "magic tee."

sketched in Fig. 92). Energy fed into leg 1 exits from both output points in phase, and therefore the two outputs (the radiations from them) add in the direction in which the outputs are pointing. A small directional beam pattern thus results which can be sharpened by directing it toward, say, a lens or parabolic reflector. For energy introduced into leg 2, however, the energy leaving the two output legs is *out of phase,* and a beam pattern having a deep dip or *null* directly ahead results. (This pattern with its null can also be sharpened by the use of a lens or reflector.)

In the radar, a single pulse (a "monopulse") can thus give not only information on the *range* of a reflecting target (through the measurement of the time it takes for the pulse to return) but also extremely accurate information on the *direction* of the target, through the process of comparing the magnitude of the received signals in legs 1 and 2. If the target is exactly on boresight, output 2 is at a minimum and 1 is at its maximum. If the target is off to one side, output 2 has a higher-than-minimum value, and output 1 is somewhat less that its maximum value.

So we see that technology benefited because observant microwave engineers *thought about* the little arrowhead on the waveguide drawing of Fig. 68 *and* because they decided to build and experiment with the strange-looking device of Fig. 92.

7

LENSES

Lenses for focusing light waves have been known since the early days of optics, with glass being the prime material for use in telescopes, prisms, and other optical epuipment used by the first astronomers. The focusing effect of a glass lens for light waves exists because of the different *velocity* which light waves experience when they leave air and enter glass.

"VELOCITY" FOCUSING

We saw in the previous chapter that when radio waves enter a metallic waveguide they too experience a change in velocity. Now the knowledge of the existence of this unusual velocity effect in a waveguide led to some off-the-beaten track thinking about possible "waveguide spinoffs," ways in which this velocity effect could be put to use. As a result, during World War II, a new kind of *lens,* a *waveguide* lens for radio waves, came into being.

The standard converging lenses made of glass are thick in the center and thin at the edges. This configuration is necessary because waves in glass and other optical refracting materials have a *lower* wave velocity than the velocity of light in air or in "free space." But since waveguides, on the other hand, produce a phase velocity *greater* than in the free-space velocity, a *convergent* lens becomes *concave* in shape when waveguide structures are used. This is shown in Fig. 93. Such an effect was obviously not known to exist when the early optics experts used the term "convex" to describe the top (glass) converging lens of Fig. 93, because "convex" is derived from the same root as "converging" so that for a *concave* lens to be *convex* would be a contradiction.

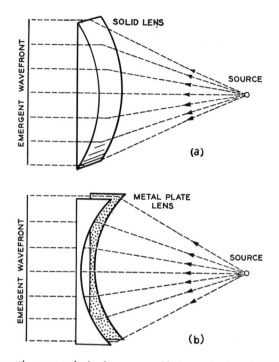

Fig. 93. *Because the wave velocity in a waveguide is greater then it is in air, a focusing waveguide lens (b) must be made concave, in contrast to a convex glass lens (a).*

To see how waveguides can refract microwaves, let us imagine a stack of rectangular waveguides cut to proper length and placed side by side as shown in Fig. 94. Waves with flat wave fronts arriving from the left emerge at the right with their wave fronts tilted because of the higher wave velocity within the guide (indicated in the figure by the

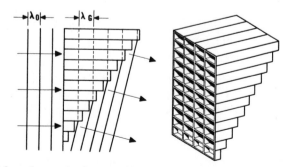

Fig. 94. *A triangular stack of waveguides (a "prism") causes microwaves to be tilted downward because of the higher wave velocity in the guides.*

longer guide wavelength λ_G). The structure behaves, for microwaves, like a prism, so that the condition of higher velocity in the guide provided a useful "spinoff" in the form of a new kind of wave "refractor." This refracting property of waveguides has been used widely in the construction of lenses of various types and sizes.

A CONSTANT-THICKNESS LENS

One way of making a waveguide lens is to have the *thickness* remain constant and the *refractive index* change over the lens aperture. This was the first method that I used. As shown in the patent sketch in Appendix 36, the height of the waveguide of Fig. 68 was gradually increased (with its width remaining constant) so as to form a *sectoral horn*. At the rectangular aperture of this horn, the arrow of Fig. 68 would be vertical, also very long, parallel to the long side of the rectangular aperture. Waves entering from a waveguide at the base of the horn would propagate to the large aperture, where their wave front (phase front) would be circular (cylindrical). At this point, a cylindrical glass lens (similar to the one in Fig. 93) could be inserted into the horn aperture so as to convert the circular wavefront to the desired *flat* one. The procedure actually selected for the "horn-lens" of Appendix Item 36, however, was to add a constant-length (metallic waveguide) section to the horn which had its width gradually narrowed from the center toward the edges. At the edges, the narrower width of the waveguide causes the waves to propagate at a higher wave velocity (phase velocity) than at the center, so that the constant-length (constant-thickness) *lens* thus transforms the circular wave front to a flat one (just as the cylindrical glass lens would).

THE METAL PLATE LENS

The success of this waveguide lens showed that such lenses or prisms need not be made of "stacks" of waveguides as in Fig. 94. In Europe, several such "stacked waveguide" lenses were constructed near the end of World War II. Secrecy procedures, as exemplified by Appendix 37, drastically restricted the flow of information during that period. The item also shows that three of the four later lens memoranda (numerous ones are not included in that list of four) were issued in 1944

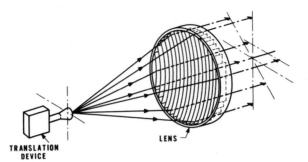

Fig. 95. A two-dimensional waveguide lens being energized by microwaves issuing from a horn at the left. From U.S. Patent No. 2,736,894.

(the numbers 44 and 45 indicate the year of issuance). The version of a waveguide lens was made as sketched in Fig. 95. The lens was 18 inches in diameter, and it was, as far as I know, the first two-dimensional waveguide lens. Successful tests of this metal-plate structure encouraged further exploration of numerous variations. Figure 95 brings out the fact that lenses making use of the increased wave velocity property of waveguides must be thick at the edge and thin at the center, the exact converse of a glass lens. Figure 96 shows a water wave ripple pattern that simulates the wave-focusing effect of a waveguide lens. As was shown in the ripple-tank pattern of the previous chapter, water waves propagating in a simulated radio waveguide situation have their wavelengths increased in length. This is equivalent to an increase in their

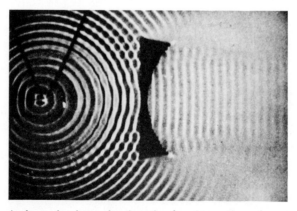

Fig. 96. A ripple tank photo showing the focusing action of a waveguide lens. Courtesy the Naval Research Laboratory.

velocity (phase velocity). The figure shows that the circular waves (originating at the center of the circles) become flat after passing through the waveguide lens. The top and bottom portions of the flat waves at the right traveled along the longest paths within the high-velocity lens section, and this was responsible for the conversion to plane waves.

STEPPED LENSES

To reduce further the weight of waveguide metal-plate lenses, the process called "stepping" was developed. In an unstepped circular lens, starting at the center of the lens and moving outward toward the rim, the thickness increases continuously. In a stepped lens, "setbacks" where the lens thickness is reduced abruptly are incorporated. Because the step design is based on one given wavelength, such lenses are feasible only in those systems that employ frequencies lying close to the design frequency. Proceeding outward from the center of the circular lens of Fig. 97, when the thickness increase equals one design wavelength as measured within the lens—that is, one guide wavelength—the lens thickness can be reduced to its original center thickness. At succeeding points, wherever the thickness increase again equals a guide wavelength, succeeding steps are introduced. Figure 98 shows a lens that has been stepped in the vertical direction only. It is called a "cylindrical lens," inasmuch as it focuses only in one plane; that is, it focuses cylindrical waves emanating from a *line* source, rather than spherical waves emanating from a point source, as in the case of the lens of Fig. 97. This cylindrical lens was incorporated in a World War II mortar-locating radar. Two line sources were placed one above the other, causing the combination of these with the lens to generate two pencil beams. The construction of the line sources was such that their contributions to the beam position (its horizontal or azimuth location) could be varied (scanned). This resulted in two almost horizontal planes, one pointed in a higher direction than the other, being continually under observance (by virtue of the horizontal or azimuth scanning process). Whenever a mortar shell, on the upward portion of its trajectory, passed through these two planes, its azimuth position in each was ascertained, and a similar determination was made on the downward portion of its trajectory. Since a shell trajectory is completely defined when a number of

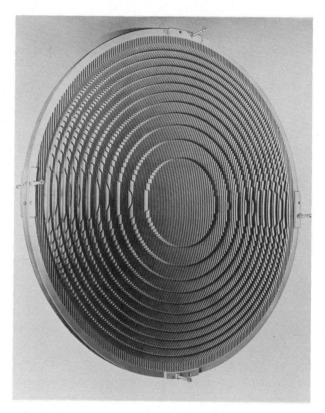

Fig. 97.   A stepped circular waveguide lens antenna designed for U.S. Navy shipboard radar use.

points on it are specified, the location of the origin of the mortar shell (relative to location of the radar) could quickly be calculated from this radar information, and action to silence the mortar could then quickly be taken. A field model of this radar, designated the "TPQ-2," was in production in December 1944.

RADIO RELAY

Shortly after World War II, the concept of communication systems using microwave radio circuits, called "radio relay circuits," began to look very promising, since they could provide much wider "roadways" for telephone and television transmission. On these, the telephone calls

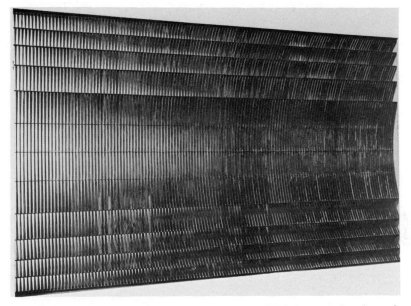

Fig. 98. A stepped cylindrical lens used in the 1944 TPQ-2 mortar-locating radar.

or television signals are sent out on a radio beam, and they reach their final destination by way of numerous *relay* stations along the way. At a relay station, the radio signal, which has traveled perhaps 20 or 25 miles, is received, amplified, and then retransmitted onward to the next station. The use of lenses for these radio beams causes an extremely small amount of interference between the strong transmitted signal and the weak received signal. The more common parabolic reflector or "dish" antenna as used in most radars would generate a significant amount of "spillover" signal from the high-power transmitting disk to the highly sensitive receiving disk. This effect is shown in Fig. 99. In some microwave antenna applications, a horn alone can be used (no lens), but for an aperture of, say, 10 feet, the length of horn required to provide an adequately plane emerging wavefront is prohibitive.

Accordingly, it was the horn-lens which was chosen for the first microwave radio relay circuit operated by the Bell System, installed between New York and Boston (Appendix 38).* A 10- by 10-foot

* W. E. Kock, "Radio lenses," *Bell Laboratories Record,* pp. 193–196 (May 1946). W. E. Kock, "Metal lens antennas," *Proc. IRE, 34,* p. 828 (1946).

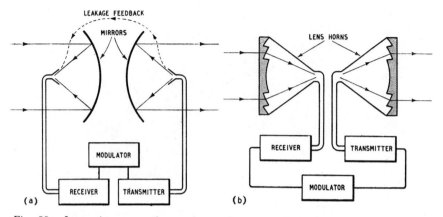

Fig. 99. In a microwave relay station, ordinary parabolic dish antennas (a) permit much of the strong transmitted signal to "feed over" into the sensitive receiving system. The shielded horn-lens antenna (b) strongly suppresses this effect. Courtesy Wireless World.

experimental model of this New York-to-Boston lens is shown in Fig. 100. Work on the New York–Boston relay was begun in 1946, and the route taken for this circuit is shown in Fig. 101. Four units were required, two for each direction of transmission, comprising a transmitting and a receiving horn-lens. Figure 102 is the cover of a Bell System brochure describing the New York–Boston relay system.

FOAMED DIELECTRIC LENSES

It was about 1946 when "foamed" dielectrics were just becoming available in quantity, and Figs. 103 and 104 show two examples of waveguide lenses in which accurately cut, lightweight foam spacers were used to support metal foil waveguide plates forming the lenses. These versions were therefore extremely lightweight ones. In the lens of Fig. 103, the copper foil plates were cut to proper shape before being affixed to the foam slabs. The lens of Fig. 104, on the other hand, was first assembled and then turned on a woodworking lathe, thus eliminating the need for cutting out beforehand each individual copper foil plate.

Thus the waveguide lens, a spinoff from the strange velocity properties inherent in the newly developed wave transmission structure, the waveguide, was able to play a part in making possible a new wideband, interstate communications network.

Fig. 100. *An experimental model of the shielded lens antennas employed in the first wide-band microwave radio relay circuit (between New York and Boston). The author is at the left.*

Fig. 101. *The route of the microwave relay circuit between New York and Boston.*

Fig. 102. The cover of a Bell brochure refers to the seven relay points of the New York–Boston circuit.

WIDER BANDWIDTH NEEDS

Many of the microwave components in the New York–Boston radio relay had a limited bandwidth capability. Initially this situation was not of great concern because the bandwidth transmission capacity of the relay circuit far exceeded that of the then existing coaxial cable. However, as rapid developments in technology caused large numbers of these microwave components to acquire broader and broader bandwidth capabilities, it became apparent that the stepped metal-plate lenses with their bandwidth limitations (as noted above) would probably not be looked upon too favorably for use in the envisioned transcontinental extensions of the New York–Boston radio relay circuits. Accordingly, a case of "necessity is the mother of invention" came into being, with an innovation *needed* and an innovation *resulting*. This development had its "birth" more than a decade before, in one of the writings of the great German scientist, Max Born, recipient of the 1954 Nobel Prize in Physics.

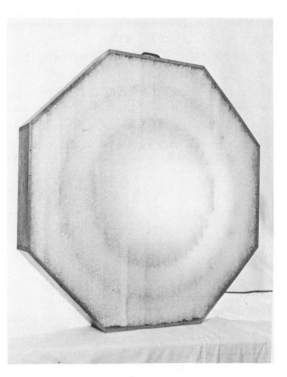

Fig. 103. Plastic foam slabs support the metal foil lens plates in this model.

Fig. 104. The contour of this foil–plastic foam lens was achieved by turning the entire unit on a woodworking lathe.

BORN'S LIGHT WAVE ANALYSIS

In his 1933 book *Optik* (one of the foremost books on optics of all time), Born presented an elegant discussion of the behavior of light waves in a continuous medium (for example, light waves passing through a glass lens). He showed that the behavior of these waves can be arrived at by assuming the material to be an assemblage of tiny, closely spaced, reradiating particles, each acted on individually by the electromagnetic action of the light waves. The action of the light waves causes these particles to reradiate the same electromagnetic waves, the same light waves, which are passing through the medium. Fifteen years after Born's book was published, a spinoff resulted in the form of a concept called "artificial" dielectrics. This concept proved useful in the field of microwave radio, permitting, among other things, the channel width (the conversation capacity) of the radio relay circuits to be increased by severalfold.

When an electromagnetic wave passes through an assemblage of discrete reradiating particles, the fluctuating charges which it induces on the particles regularly reradiate energy, and this energy adds to or subtracts from the wave itself. If there are many particles per wavelength, the reradiated energy combines with that of the wave which excites it so smoothly that the wave merely suffers an alteration in velocity. For light, for which the wavelengths are so short, particles of molecular dimensions are required for such a behavior to exist. It is well known that transparent crystals are actually regular arrays of extremely tiny particles called molecules. Glass is not composed of a regular array of molecules, but if the arrangement of these particles in a substance is adequately uniform in density and composition, a "smooth" refraction of light will take place if there are many molecules per wavelength. With this knowledge that a dielectric substance behaves as it does because it is composed of discrete particles which are excited by the electromagnetic waves, we ask ourselves the question: Might it not be possible to simulate this behavior *artificially*; that is, can the behavior of natural dielectrics be simulated by forming an array of conducting particles on a scale very large, of course, compared with that in a natural dielectric but still small compared with the wavelength of the radiation to be used?

THE FIRST ARTIFICIAL DIELECTRICS

It was this simulation which I believed could furnish a possible solution to the radio relay problem of providing wider bandwidth lenses without sacrificing the weight advantage of waveguide lenses (as compared with 10-foot glass lenses). So I asked myself, "What are the most satisfactory structures for such artificial dielectrics?" Since all substances are composed of assemblages of very tiny particles (atoms and molecules) and since these particles or groups of particles have electric charges, they are affected by the presence of an electric field. Thus if we place a tiny particle of material between the plates of a charged electric condenser, we find that the particle becomes strained or distorted. The negatively charged units in the particle are attracted by the positive plate of the condenser and *vice versa*. If we reverse the polarity of voltage applied to the condenser plates, the particle is strained in the reverse direction.

Now electromagnetic waves (light waves or microwaves) establish electric fields similar to the electric field existing between the plates of a charged condenser. In the wave case, however, the polarity of the electric field continually reverses itself, at a rate that depends on the frequency of the electromagnetic wave. Thus for our 60-cycle house current the electric current is alternating at a rate of 60 alternations per second and the electric field it would create across the plates of a condenser would change direction 60 times a second. For *microwaves* propagating within a waveguide, the top and bottom walls of the guide act like the parallel plates of a condenser. In this case, the electric field between the "plates" is alternating or reversing itself at the rate of several thousand million times a second. This rapid field reversal continues to exist even after the waves emerge from the waveguide and are propagating in free space. A similar but even faster field reversal is created in space by light waves. One of the tiny particles of Fig. 105, if placed in the path of waves of violet light, would experience an electric field reversal 7×10^{14}, or seven hundred trillion, times a second. Because every action has an equal and opposite reaction, the rapid reversals of strain on the particle react on the light wave which caused it, and this reduces its velocity of propagation. Thus light moves through glass at about two-thirds of its velocity in air or vacuum. Lower-frequency electromagnetic waves, such as microwaves, experience this same effect, and glass accordingly exhibits the same refractive

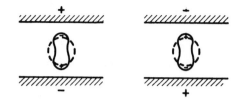

Fig. 105. The charged particles of a molecule of glass or dielectric are attracted to the oppositely charged plates of a condenser. Reversal of the electric field (at the right) reverses the direction of stress on the molecules.

index (or wave velocity reduction) for microwaves as it does for light waves.

Since the wavelengths of microwaves are specified in inches, as compared to millionths of an inch for visible light, an obvious question follows. Cannot the particles involved in Max Born's wave refractive process be made much larger if we are dealing with microwaves rather than light waves? An affirmative answer appeared likely, and accordingly the first artificial dielectrics were put together.

SPHERE AND DISK DIELECTRICS

It was decided that the first particles to be tried would be conducting spheres, since it seemed likely that between the plates of a condenser (Fig. 106) such spheres would act very much like the molecules of a true dielectric. To simulate the molecular crystal lattice properly, the conducting spheres had to be spaced apart and insulated from one another. For a quick and easy experiment, I purchased a string of imitation pearl beads, coated them with conducting paint, and mounted them on thin (insulating) cylindrical wooden sticks ("kite sticks"). This first artificial dielectric lens is shown in Fig. 107. Because the number of spheres in

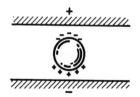

Fig. 106. When a conducting sphere is in an electric field, the free electric charges on the surface of the sphere cause it to behave like the molecule of Fig. 105.

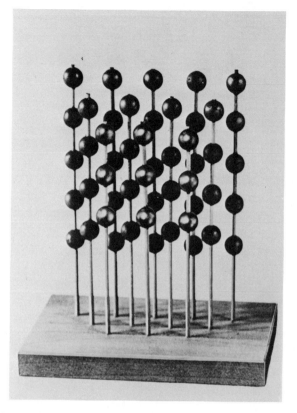

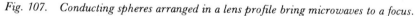

Fig. 107.   Conducting spheres arranged in a lens profile bring microwaves to a focus.

this lens was small, the ideal lenticular shape of the assemblage could only be approximated. However, the focusing action was quite pronounced. It was then recognized that if disks were substituted for the spheres a weight advantage would result, and if they were mounted so as to present their maximum area to the incoming wave an effect similar to that of spheres should be observed. A further reduction in weight was achieved by making the lens disks of thin, lightweight metal foil. Figure 108 shows a lens with copper foil disks supported on circular polystyrene foam slabs. Figure 109 shows this lens forming a beam of microwaves. Another version of a lens using foam slabs to support the planar refracting elements is shown in Fig. 110. Here molten tin was sprayed through a mask onto the circular foam slabs, where it solidified as shown.

Fig. 108. Copper foil disks affixed to circular foam slabs provide a very lightweight microwave lens.

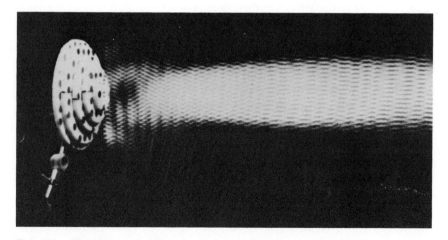

Fig. 109. The disk-array microwave lens concentrating electromagnetic wave energy (arriving from the left) into a narrow pencil-shaped beam.

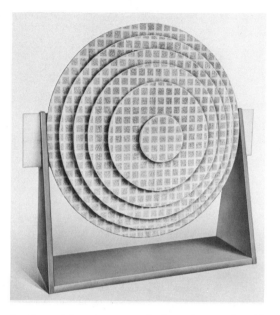

Fig. 110.   *Circular foam slabs were sprayed, through a mask, with molten metal. It solidified into the pattern shown.*

STRIP LENSES

The third variety tried was the strip lens. If we imagine the disks of the lens of Fig. 108 joined together along horizontal lines, we arrive at the strip lens of Fig. 111. Very large strip lenses can use foamed plastic to support the thin metal strips. This procedure is sketched in Fig. 112, and Appendix 39 shows my colleague William Legg standing behind a partially assembled experimental model of such a lens.* Appendix 40 shows the wide interest evoked by the announcement of this lens (Appendix 41).

THE TRANSCONTINETAL RELAY

When the Bell Telephone System's New York–Boston microwave circuit was extended to transcontinental use, such foam-supported strip lenses were selected because of their much broader bandwidths relative

* W. E. Kock, *Bell Syst. Tech. J.*, *27*, p. 58 (1948).

Fig. 111.   The disks of the lens of Fig. 108 can be replaced with horizontal strips.

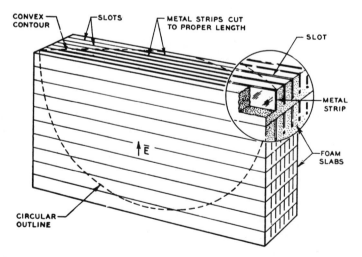

Fig. 112.   A larger form of the lens of Fig. 111 in which plastic foam is used to support the metal strips.

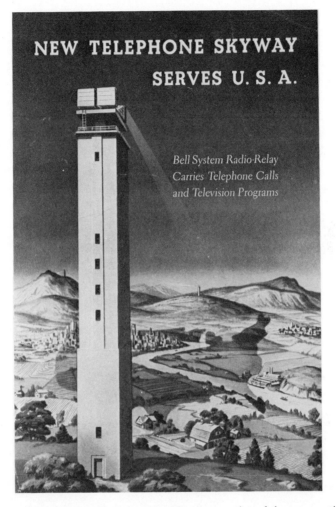

Fig. 113. *Because of the flat terrain in the midwestern section of the transcontinental relay system, the towers must be quite tall. This photo is from the cover of a Bell brochure describing this relay circuit.*

to the earlier waveguide lenses. One of the cross-continental relay system's towers is shown in Fig. 113; three of the four horn-lens antennas of the kind used in each two-way relay are visible. Because of the flat terrain of the midwestern United States (no mountains), the towers had to be much taller than in the New York route using these "artificial dielectric" lenses. The New York–Chicago extension horn-lenses went into commercial operation in 1950.

THE BELL TRANSCONTINENTAL CIRCUIT

Later, this relay circuit was extended across the U.S. continent; the route selected is shown in Fig. 114. First television programs and phone calls were carried over the circuit on August 17, 1951; Appendix 42 shows Wayne Coy, Chairman of the Federal Communications Commission, and A. T. Killingsworth, Vice-President, and Cleo Craig, President, both of the AT & T, opening the system. On September 4, 1951, President Truman, addressing the opening session of the Peace Treaty Conference in San Francisco (Appendix 43) could be seen as well as heard by millions of Americans from the Pacific to the Atlantic by virtue of this new radio relay system. The Bell Telephone System Almanac for 1952 featured this coast-to-coast network as its cover (Appendix 44).

The October 1970 *Bell Laboratories Record,* in its "20 Years Ago" section, referred to developments in the "TD-2" network as follows:

MICROWAVE NOTES

September saw the completion of four new TD-2 microwave radio relay systems for transmitting television. Two 380-mile channels between Los

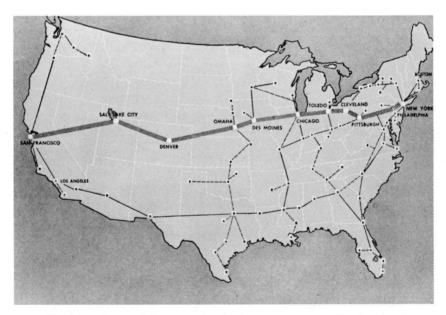

Fig. 114. A map of the route taken by the coast-to-coast radio relay circuit.

Angeles and San Francisco went into service on September 15; one is a northbound channel and the other, although normally southbound, is reversible. Because of the height of some of the repeater sites along this route, two of the hops are 65 miles long—unusually long for a microwave system. On September 18, commercial television was established over a 315-mile route between New York and Washington. It uses stations of the system between New York and Chicago as far as Clark's Knob, some 40 miles west of Harrisburg. From here a branch runs to Garden City, Virginia, where a temporary radio link extends the system to Washington. One northbound and one southbound channel are provided. The other two systems consist of a 110-mile two-channel system between Dayton and Indianapolis and a 455-mile section of the transcontinental system between Chicago and Omaha. The latter system has two channels, westbound only, with spur circuits to join Davenport and Ames, Iowa. Both of these systems went into service on September 30.

Similarly, in the September 1975 issue of the same journal (in the "25 Years Ago" section) the following appeared:

A FOCUS ON BETTER, LOW COST TELEPHONE SERVICE
In the new microwave radio relay system between New York and Chicago, giant lenses shape and aim the wave energy as a searchlight aims a light beam. Reasoning from the action of molecules in a glass lens which focuses light waves, Bell Laboratories scientists focus a broad band of microwaves by means of an array of metal strips. To support the strips these scientists embedded them in foam plastic which is rigid, light in weight, and virtually transparent to microwaves.
This unique lens receives waves from a waveguide at the back of the horn. As they pass across the strips, the waves are bent inward, or focused, to form a beam like a spotlight. A similar antenna at the next relay station receives the waves and directs them into a waveguide for transmission to amplifiers. This new lens will help to carry still more television and telephone service over longer distances by microwaves.

HIGH REFRACTIVE POWER LENSES

An interesting extension of the artificial dielectric concept was the fabrication of a metal-strip microwave lens having a far higher refractive power than possessed by any known natural material. This was accomplished by placing the metal strips very close together, making them out of copper foil, and mounting them on thin transparent plastic sheets. Also, the metal strips were made to overlap, with the plastic sheets furnishing the needed insulation to prevent adjacent ones from "short-circuiting" the electric field of the microwaves. This lens, shown

Fig. 115.   Conducting strips placed extremely close together form a refractive medium having an effective dielectric constant of 225.

in Fig. 115, exhibited a refractive power (refractive index) which was 10 times that of glass.

We have thus seen that an elegant explanation of waves in crystals by a Nobel Laureate in physics led to several new forms of radio lenses which then helped to make cross-country television possible.

PATH LENGTH LENSES

Another lightweight microwave lens followed the waveguide lens and the artificial dielectric lens. It is called the "path length lens" because its operation is based on the control of the travel times by making the lengths of the focused waves equal. In this lens, the wave travel time is controlled by forcing the waves to travel longer paths at their normal free-space velocity, as contrasted to a glass lens where the wave velocity is reduced.

Figure 116 shows a cross-section of a lens structure which accomplishes this desired result. The tilted straight lines within the lens contour are cross-sections of metal plates extending into the paper and

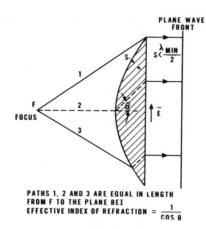

Fig. 116. When the path of waves is controlled, all rays can be made to have the same path length from focus to plane wave front.

out for the full width of the lens. This tilted flat plate structure is illustrated in Fig. 117, a photo of the first of this type of lens. As seen in Fig. 116, the *path lengths* of rays 1, 2, and 3 are equal, so that even though the waves pass through the lens with the same velocity as they have in free space (outside the lens), the lengths of their (tilted) paths

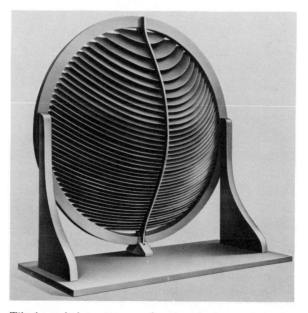

Fig. 117. Tilted metal plates can cause focusing of microwaves (and sound waves).

within the lens, when added to their additional path lengths from the focus to the lens, are the same, and the waves emerge having the desired plane wave front. As indicated in Fig. 116, their electric vector **E** is perpendicular to the plates, so, in contrast to the waveguide lens, the plates cause no change in the wave velocity.

Also in contrast to the waveguide lens, the path length lens has a very wide bandwidth capability, and because of this, and because of its relatively simple construction, it has been employed in microwave relay applications. Figure 118 shows a closeup of such a lens designed in

Fig. 118.   Top: A Japanese path length lens. Bottom: These lenses used in a microwave radio relay tower. Courtesy Mitsubishi Electric.

Japan for relay use and also a relay installation of four such horn-lens antennas. The application date of the U.S. patent (Appendix 45) shows that the path length lens followed closely the waveguide and artificial dielectric lenses.

LENSES FOR LOUDSPEAKERS

About the time of the commercial introduction of microwave lenses into the Bell System's radio relay circuits, my colleague and close associate F. K. Harvey and I began thinking of applying the sphere and disk lenses of microwaves to acoustics. Harvey looked upon these elements as "obstacles," and in this "off-the-beaten-track" approach various forms of "obstacle" lenses were developed for focusing sound waves. Soon he and I were successfully conducting experiments showing that some of the *already constructed* microwave lenses did indeed focus sound waves, and we were also designing and constructing *new* models.

Figure 119 is an "obstacle table," prepared by Harvey, giving formulas of "acoustic dielectrics." Because sound waves are not polarized, the formulas differ from those specifying the microwave dielectric designs and Harvey's obstacle concept was therefore very useful.

Now the testing of microwave lenses can be rather conveniently carried out using fairly simple outdoor test ranges. Not so, however, for acoustic devices. Tests outdoors are likely to be affected by unwanted noise, and the walls of the usual indoor room strongly *reflect* sound, and this likewise affects the test results. Fortunately, at the Murray Hill Laboratories of Bell there is a very large anechoic or "free-space" room, whose walls are covered with huge amounts of sound-absorbing material to eliminate acoustic reflections. This room is one of the quietest and most nonreflective ones of its kind. Appendix 46 shows this room during the television show of the popular newscaster Edward R. Murrow. This broadcast followed national election day in 1952, and the first portion of Murrow's program included many very noisy scenes, showing loudly cheering supporters of successful candidates. For contrast, Mr. Murrow closed his program with a few moments at what he called "the quietest spot on earth." It was in this room that Harvey conducted many of the tests to determine how effective "obstacle lenses" would be for sound waves.

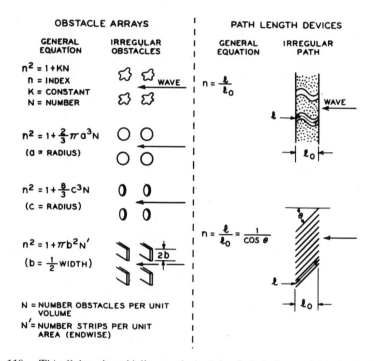

Fig. 119. This "obstacle table" was devised by Bell Laboratories scientist F. K. Harvey. It presents the various formulas needed to achieve the desired "refractive" properties of obstacles.

MICROWAVE LENSES FOR SOUND

In selecting earlier microwave lenses for such tests, it was evident that certain structures, such as the one shown in Fig. 108, cannot properly affect sound waves because the sound waves will be reflected by the foam. However, if the disk lens has an open structure, sound waves can at least pass *through* it, and *possibly* be affected by the disks. Figure 120 shows such a lens being tested in the Murray Hill anechoic room (*an*echoic, meaning that the walls are so absorptive that no *echoes* can be generated), and Fig. 121 shows the same lens generating a beam of sound. Sound focusing by microwave lenses had become a reality!

But the diving into the woods continued. We noted earlier that one problem associated with loudspeakers (as used in fine radio receivers and high-fidelity reproducing consoles) is that when high frequencies are radiated they tend to be directed in a beam of sound, rather than

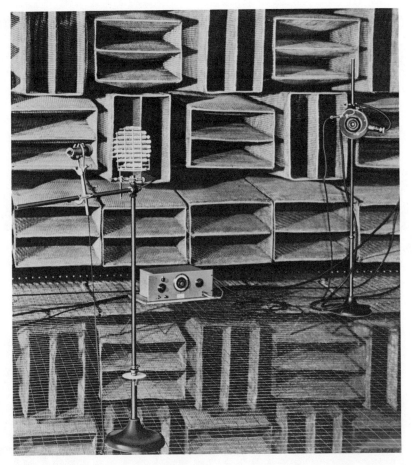

Fig. 120.   A disk lens under test in the Bell "quiet room."

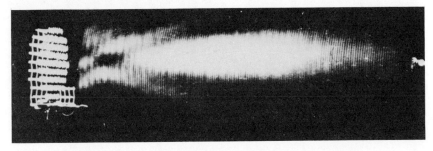

Fig. 121.   The lens of Fig. 120 focusing a beam of sound waves.

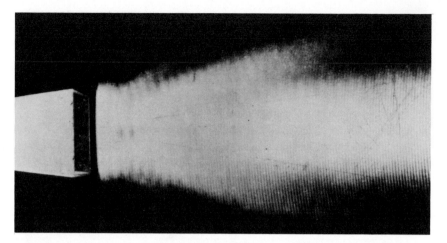

Fig. 122. High-pitched sound waves issuing from a 6-inch square horn are quite directional.

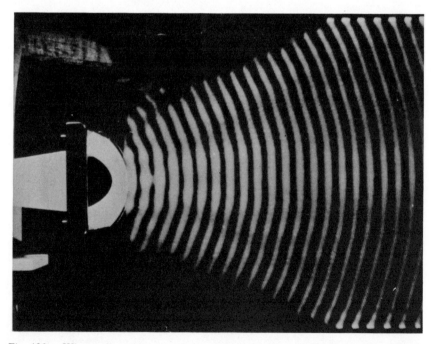

Fig. 123. When a diverging acoustic lens is placed in front of the horn of Fig. 122, the waves spread out (this photo shows the wave pattern).

being spread out equally to all parts of the room as the low frequencies are. The off-the-beaten-track thinking here went as follows: Since optical lenses can be either of the focusing (converging) variety or of the spreading (diverging) variety, it should also be possible to design acoustic lenses which can cause sound waves passing through them to *diverge* (and thus fill the room more effectively).

Figures 122 and 123 show that this effect is indeed achievable. The first photo portrays the beamed sound pattern issuing from an 6-inch-square loudspeaker, and the second, which also portrays the *wave* pattern, shows how an acoustic lens placed before that speaker causes the sound to spread out or diverge. That lens was a path length ("slant plate") obstacle lens. A commercial loudspeaker which employed a diverging acoustic lens was offered to the public a few years after publication of the original paper on such lenses by Harvey and me.* Appendix 47 shows Harvey's picture on a journal cover, and indicates the extent of the popular interest in sound wave focusing. Also, as shown in Appendix 48, some science writers, well before the advent of the commercial "Executive Phone" (distant-talking phone), were predicting the use of such acoustic lenses in "no hands" telephones! The Bell System also suggested that the recognition of the focusing effect on sound waves could be important in the future to "your telephone system" (Appendix 49). A similar advertisement, which appeared earlier when the strip lens was used in the transcontinental radio relay circuit, is shown in Appendix 50.

Max Born's elegant theory of the propagation of light in solids had been extended to the realm of acoustics, with improved loudspeakers resulting.

---

* W. E. Kock and F. K. Harvey, "Refracting sound waves," *J. Acoust. Soc. Am., 21,* pp. 471–481 (1949).

COMMUNICATIONS SATELLITES

In this chapter, we review a development which could be considered a case of serendipity in innovation. This is because the rockets required to place our present stationary satellites in orbit were actually developed partly for the space program and partly for ballistic missile use, so that the rocket developers had not really been seeking the benefits which later accrued from our present-day technology of stationary satellites. To trace the development of this very important field, let us go back a bit in time.

THE FIRST ROCKETS

During World War II, following their occupation of the coast of France opposite England, the Germans were in a position to bomb London with moderately long-range air weapons. Their first version of such a weapon was the V-1, familiarly known as the "buzz bomb." It was an unmanned aerodynamic vehicle whose motive power could be set to turn off at a certain point in its flight, causing the vehicle, with its bomb, to drop fairly close to a predesignated target, usually located in the city of London. Although the V-1 came as a surprise, the Allies were soon successful in countering this threat fairly well; it was an aerodynamic vehicle, "flying" like an aircraft, and it was hence vulnerable to standard antiaircraft fire.

Meanwhile, at a research and development center near the city of Peenemuende in the northeastern part of Germany, a far more significant development was taking place, an advanced rocket project, involving, among others, the German scientist Wernher von Braun. (Von Braun came to the United States after the war, and became an outstanding contributor to the U.S. rocket development programs, first in the U.S. Army program, later in the U.S. Air Force intercontinental missile program, and later in the NASA space program.) During the war, the Peenemuende group was working on what later became the V-2 weapon, a weapon destined to demonstrate convincingly, to unbelievers the world over, the importance of the much earlier—yet until then almost completely unheralded—efforts of the U.S. rocket pioneer, Robert H. Goddard. The V-2 was powered by an advanced form of the same rockets which Goddard had experimented with. And because the V-2 used this form of power, its trajectory closely resembled that of an artillery shell except for the much larger range. The significance of this lay in the fact that the V-2 bomb was completely invulnerable to antiaircraft defenses. To destroy *it* in flight was like trying to hit an oncoming bullet.

THE INTERCONTINENTAL MISSILE

The V-2 arrived too late to seriously influence the course of the war, but its effect on world thinking persisted. The visionary rocketeers began thinking, not of rockets merely able to cross the English Channel, like the V-2, but of rockets capable of spanning the oceans—in short, rockets to power intercontinental missiles. Because their trajectories would also be similar, except for their far greater range, to those of artillery projectiles (for which the term "ballistic" is used) weapons of this sort were given the name "intercontinental ballistic missiles" (ICBMs), and because they would behave like bullets the countering of them in flight would be extremely difficult.

Accordingly, it was not long before both of the United States and the Soviet Union embarked on programs aimed, eventually, at achieving an intercontinental missile having a nuclear warhead. The first step in the U.S. program was simply that of trying to duplicate the V-2, and, because some portions of the original German V-2's were available, the first successful launch of an equivalent V-2 (*Newsweek* called it a "German-built but American-improved" V-2) occurred shortly after the war, on the night of December 17, 1946. The programs of both countries continued to move ahead rapidly, but when the Soviet Union announced, in August 1957, a successful test of a truly long-range (intercontinental) missile, many U.S. scientists expressed disbelief.

THE REENTRY PROBLEM

One very serious and very important problem in such long-range missiles was the so-called reentry problem. These missiles not only travel a very great distance but also reach a very high altitude at the midpoint of their trajectory. They actually leave the earth's atmosphere on their flight and must therefore reenter the atmosphere. Ordinary objects traveling at the great velocity of such missiles would, upon entering the atmosphere, be burned up completely, like the majority of meteorites or "falling stars" we sometimes observe on a clear night. Accordingly, the Soviet announcement elicited skepticism from many, including several top U.S. scientists then involved in a Summer Study at Wood's Hole, Massachusetts (in which I also participated, Appendix 51). This was a National Academy of Sciences Study involving secret sessions for the U.S. Air Force (the Air Research and Development Command). During the summer, the study group was visited by many top-ranking officials of the Air Force, including General James H. Doolittle.

This group's skepticism of the accuracy of the Soviet announcement was reported in a front-page article in the August 29, 1957, *Cape Cod Standard Times,* entitled "Wood's Hole Scientists Doubt Red Claims of Successful Intercontinental Missile." It included statements (Appendix 52) such as "[The] Kremlin announcement . . . was greeted with skepticism by many of the 50 scientists who compose the National Academy of Science's Summer study group" and "One of the most vexing problems in manufacturing a workable intercontinental missile is to find a metal or plastic that can withstand the terrific heat that will be generated as the missile reenters the earth's atmosphere at 15,000 miles an hour when nearing its target. Unless we can find a material that will stay solid under these (reentry) conditions, the ICBM will burn up like a meteor as soon as it reenters thicker air near the end of its flight."

As we all know, solutions were found for the reentry problem, and the intercontinental ballistic missile now is almost a commonplace item. However, the solutions for reentry, and the warhead design procedures, were different for the two countries, and these differences led to sizably different rocket development programs. Because of the particular warhead design and reentry solution worked out by designers of the United States ICBMs, the weight of the "payload" required to be delivered by U.S. rockets was rather modest, and, accordingly, only modest-thrust rockets had to be developed. For *its* ICBM, the Soviet Union had to develop much more powerful rockets, and this situation led, naturally, to a smug feeling among U.S. ICBM engineers and scientists, a smugness that was destined to be short-lived.

SPUTNIK

On October 4, 1957, the world was startled by the Soviet announcement of its successful launching and orbiting of Sputnik I, the world's first manmade moon (manmade satellite). As Appendix 53 indicates, this was front-page news, with the newspaper articles having provocative titles such as "Russia Wins 'Baby Moon' Race, Gaining Great Popular Prestige." Also, many U.S. engineers and scientists came forth with statements such as "[The] U.S. lost the race because the Soviet Union got ahead of us in training scientists" and "Our failure to get a satellite aloft by this time is due to the lack of support given to education in the fields of engineering and science."

But other newspaper articles, such as the following by Alton Blakeslee, accurately described the difficulties in orbiting the baby moon:

> The Soviet success means that they have achieved a most difficult task in rocketry. Just shooting a rocket hundreds of miles high is not the problem. It is to make the multistage rocket turn in its flight to be parallel to the earth. Then the last rocket kicks the moon forward in a great surge so its speed counterbalances the pull of gravity, and it stays in orbit. This requires the most difficult kind of rocket guidance and controls.*

The Soviet satellite was able to convince the world of its existence by its periodic radio transmissions of audible "beeps." As the newspapers noted: "RCA Communications in New York reported hearing the eerie beep beep sound from the satellite last night. The National Broadcasting Company and the Columbia Broadcasting System inter-

---

* Alton L. Blakeslee, Associated Press Report, New York (October 5, 1957).

rupted programs to broadcast the sound." Because the baby moon continued to circle the earth for many weeks, large numbers of Americans were able to obtain (nighttime) views of the "moving star" and to tune their radios so as to be able to hear the beep tone as it passed by.

INITIAL U.S. REACTIONS TO SPUTNIK

The Soviets' accomplishment came about largely because of their earlier developments of more powerful rockets for their ICBMs. As stated in an important reference work,* "The USSR's large rockets were necessitated by their large and heavy (that is, less advanced) nuclear warheads." The U.S. rockets, although adequate for U.S. missiles, were inadequate at that point for the task of orbiting a satellite. Approximately 1 month after Sputnik, President Eisenhower, to allay fears about a possible "missile gap," revealed that the United States possessed reentry capability. Development programs for more powerful rockets had been under way, but in order to gain time and speed up this development, rocket expert Wernher von Braun was called on by President Eisenhower to embark on a crash program. It is quite likely that public opinion was responsible for this action, as was also the desire to show the world that U.S. technology was fully capable of meeting the challenge. An indication of public opnion is that the then governor of Michigan, G. Mennen Williams, composed a poem on the subject (Appendix 54).

EXPLORER I

Von Braun's team *did* meet the challenge, and within less than 4 months (January 31, 1958) a U.S. "moon," Explorer I, was orbiting the earth. This accomplishment was also front-page news (Appendix 55), with Von Braun naturally sharing the spotlight. President Eisenhower, who was in Augusta, Georgia, for a weekend of golf at the time, received the good word that an Army Jupiter-C rocket had placed the United States' first satellite into orbit, and the first official

* *An Administrative History of NASA, 1958–1963,* NASA Information Division (1966).

announcement was put out that the baby moon was racing around the globe.

News reports from London stated that the successful launch "lifted spirits in Western Europe," and a West German government official in Bonn said, "The American people are to be congratulated on this great scientific achievement." Von Braun explained that the Jupiter-C was a four-stage rocket with the first stage a modified Redstone missile, one modification being that the fuel was a mixture of liquified oxygen and a "hush-hush 'rather exotic fuel.'" The newspaper report stated: "The great blast of orange flame that marked the Jupiter-C departure from the earth indicated it was one of the most powerful rockets ever launched here. Taking off more quickly than other big missiles, it gained momentum swiftly as its mighty engine thrust it high into the starry night sky. Seven minutes after the blastoff, its satellite was in orbit."

Shortly thereafter, on March 5, Eisenhower released a memo-randum recommending that the National Advisory Committee for Aeronautics (NACA) "should be renamed the National Aeronautics and Space Administration (NASA) and requesting that appropriate legislation be enacted and supplemental appropriations be provided."* A few months later, Congress passed the "National Aeronautics and Space Act," and it was signed into law by President Eisenhower on July 29, 1958.

On March 17, 1958, the U.S. Navy launched into orbit Vanguard I. It transmitted, by way of beeps, data indicating the "pear-shaped" (rather than perfectly spherical) nature of the earth. On its fourth anniversary, it was the oldest satellite, and it was still beeping. It had passed over Fort Monmouth, New Jersey, tracking station 15,712 times and had traveled 543,195,264 miles, with its orbit practically identical to its original one. It continued to transmit for over 2 more years (until May 1964).

But the bigger, earlier Soviet rockets continued to steal the show. Even before Explorer I had been launched, Russia had orbited, on November 3, 1957, Sputnik II, weighing 1,100 pounds (compared to the 30-pound Explorer) and carrying as a payload a live dog and various scientific instruments. On April 12, 1961, aboard a spaceship named *Vostok I*, Soviet cosmonaut Yuri Gagarin became the first man

* *Ibid.*

to orbit the earth. A few weeks later (May 5, 1961), in a suborbital space flight, U.S. astronaut Alan Shepard also traveled into outer space in a bulletlike (ballistic) trajectory from Cape Canaveral to the South Atlantic, and soon thereafter (February 20, 1962) U.S. astronaut John Glenn, on the spaceship *Friendship 7*, became the first American to orbit the earth (Appendix 56). The blastoff was observed on television by millions of Americans, including President Kennedy, who had been elected in 1960.

The U.S. space program soon reached full momentum, with newer, bigger, and more powerful rockets appearing almost monthly. Thus, on June 3, 1965, astronaut Edward White took a 21-minute "walk in space" (Appendix Item 57) while his partner James McDivitt remained in the orbiting spaceship *Gemini 4*. The spaceship orbited the earth 66 times. This great speed of development was a requisite if President Kennedy's announced goal of landing a man on the moon by 1970 was to be achieved.

APOLLO 11

America's goal in space was reached when U.S. astronaut Neil Armstrong became the first man to set foot on the moon on July 20, 1969 (Appendix 58). This historic event involved spacecraft number 11 of the Apollo series. Because of my involvement in setting up NASA's Electronics Research Center in Cambridge, Massachusetts (I had been loaned by the Bendix Corporation for 2 years for this task), my wife and I were invited to witness, on July 16, the launch of Apollo 11 (Appendix 59). Again, many saw the blastoff on television, and it is estimated that about 530 million people watching television sets around the earth also saw, televised from the moon, Neil Armstrong's image and heard his voice as he took "one small step for man, one giant leap for mankind." While Armstrong and astronaut Buzz Aldrin were on the moon, having landed with that portion of the spaceship called the "Lunar Module," the third astronaut, Michael Collins, remained in the main spacecraft, circling the moon. Figure 124 shows Apollo 11 astronaut Buzz Aldrin deploying one of the pieces of scientific equipment on its proper spot on the moon.

With the successful splashdown of Apollo 11, the world's image of the United States as a technological leader, tarnished by Sputnik, had

Fig. 124. Apollo 11 astronaut deploying a seismometer. The LM Eagle is in the background.

again become bright, and the dreams of the U.S. pioneer Robert Goddard had come true. For Goddard, in an interview in the *American Weekly* in 1921, had declared that "the only way to convince those who are skeptical of a rocket reaching the moon is to show them."* And NASA did.

COMMUNICATIONS

The continued growth of the world's economy over the years has made it desirable and necessary that more and more people be able to communicate with one another, and so the history of communications technology is replete with examples of successful efforts aimed at increasing the useful bandwidth (the width of the frequency band required to transmit a given type of signal) of communications circuits. The proven value of telegraph messages (dots and dashes at quite slow repetition rates with a bandwidth requirement of only a few hundred hertz) led to the development of the telephone, which even today

* Reported in the 60th Anniversary Issue of *American Weekly*, p. 26 (October 20, 1956).

requires, for effective voice transmission, a bandwidth of only 4000 hertz.

A significant advance occurred with the development of the much broader-bandwidth "coaxial cable," invented by Lloyd Espenschied of the Bell Telephone Laboratories. This cable, along with the carrier system, whereby many phone circuits are stacked on top of each other, eventually enabled hundreds of thousands of phone conversations to be transmitted over a single circuit. Then, at the 1939 World's Fair at New York City, U.S. television was shown widely for the first time. It was obvious that network television, comparable to the network radio of many years' use, would become a highly desirable and very important possibility, and the coaxial cable did provide some help in this field.

As we saw in the last chapter, the next step came in the development of microwave radio relay circuits. The relay stations spanned the United States with bandwidths far exceeding those of the then-existing coaxial cables, permitting network television to provide color TV pictures to much of the nation. These circuits utilize tall towers because of the need for establishing a "line of sight" for the very-high-frequency radio waves. The earth's curvature thus places a limit on the separation of two tower-borne microwave antennas, with the possible separation being larger the higher the towers. Accordingly, although the Bell System's first transcontinental microwave radio relay developments were extremely important for overland routes, they could not be used in spanning the oceans. For such applications, the older, undersea coaxial cables were the only possibility.

COMMUNICATIONS AND THE SPACE PROGRAM

Quite some time before the historic moon landing event took place, serendipity had crept into the space picture, leading to worldwide satellite communications. To set this valuable space spinoff in proper perspective, it should be noted that the use of satellites for communication purposes had been suggested, long before Sputnik, by the British scientist Arthur C. Clarke in an article in *Wireless World*.* (Clarke did note that rockets to achieve orbital or escape velocities were still some years away.)

* A. C. Clarke, "Extra-terrestrial Relays," *Wireless World*, pp. 305–308 (October 1945).

Fig. 125. This sketch, published in 1961, predicted transatlantic television transmission via satellite.

A satellite is uniquely qualified, by its line-of-sight feature, to be equivalent to a radio relay tower many thousands of miles high. It fulfills therefore the biggest difficulty of wideband microwave communications. Because of the earth's curvature, these signals are now relayed by radio relay towers spaced every 30 or 40 miles. It is obvious that this requirement limits this form of transmission to developed land areas such as those of the United States. The availability, in a satellite, of several orders of magnitude increment in relay tower height (as suggested by a sketch which appeared in 1961, Fig. 125) is without question a very significant advance in global communications technology. Accordingly, one of the most important impacts of our *space* program is found in the field of communications.

ORBITING SATELLITES

Just as the early space experiments such as the Sputniks and the Explorers involved satellites orbiting every 2 hours or so, so the first use of space for communications used similar rapidly orbiting systems.

Score, launched in 1958, transmitted taped messages, and Echo I, the first *passive* communications satellite, relayed voice and television signals by *reflecting* them off its large spherical area.

The orbiting satellite which made the biggest headlines, however, was Telstar (Appendix 60), the Bell System's experimental communications satellite, looked on at that time by Bell as a step "toward the eventual goal of achieving continuous broadband communication across oceans by way of microwave radio."* As the *New York Herald-Tribune* account noted, "Before Telstar, relaying a live TV signal from the U.S. to Europe would have required a mid-Atlantic tower an impossible 145 miles high."

The Telstar satellite, launched by NASA but with costs paid by the Bell System, was a 34-inch-diameter sphere. A signal beamed by one ground station at the (moving) satellite was picked up, amplified, shifted in frequency, and retransmitted by the satellite to the "across-the-Atlantic" ground station. Because of the limited size of the satellite, the power of the signal retransmitted was only 2 1/2 watts, spreading out uniformly in almost all directions. To scoop up as much as possible of this radiation, very much weakened by its 3000-mile trip through space, Bell Laboratories engineers designed the largest horn-antenna of its time. Its 3600-square-foot opening received about a billionth watt of the broadband signal; its sides kept out unwanted ground radiation. This mammoth horn (at Andover, Maine) imposed requirements more exacting than those for any other structure of its size ever built, for it had to track its tiny target smoothly and continuously, to an accuracy of better than 1/20 degree. Design engineers had to consider that the weight of the antenna itself would introduce bending, and a different amount of bending for every position it took. It was therefore built as rigidly as possible and—for its size—more accurately than a fine watch. A 70-foot-diameter rotating ring gear, for example, was machined to a tolerance of less than 1/32 inch. To remove the factors of wind stress, icing, and rapid temperature changes, the entire horn was covered with an inflated radome, 210 feet in diameter and 161 feet high.†

The Telstar satellite successfully relayed color television programs between Europe and the United States, and, in 1967, the Soviet Union decided on that orbiting satellite technique for implementing broadband

* Tenth anniversary of Telstar," *Bell Laboratories Record,* pp. 202–203 (June–July 1972).
† *Ibid.*

communication to about two dozen cities which are quite distant from Moscow, including several in Siberia, with a system called "Orbita Molina." This system was still in use in the Soviet Union in 1974.*

GEOSTATIONARY SATELLITES

Because Telstar and other orbiting satellites, such as Score, Echo, Courier, and Relay, orbited the earth once every 2 hours or so, at altitudes of about 100 miles above the earth's surface, the ground stations, in order to send to the passing communications satellites (and to receive from them) the radio communication signals, had to be equipped so that the aiming angle of their directional beams could be continuously varied (so as to always be pointing at the orbiting satellite). Accordingly, with the communications satellites doing an indifferent job in their low-orbit activity, chiefly because of the difficulty of maintaining contact with them, it was apparent that something better was needed. That something was a breakthrough, a communications satellite which was to demonstrate the greatest promise for worldwide satellite communications.

It appears that Dr. Harold A. Rosen and his associates at Hughes Aircraft were the first to consider theoretically (and then convert their creative concept to one of the great innovations of our time) the world's first synchronous-orbit satellite. They envisioned that a vehicle could be launched and made to orbit at a specific distance above the earth and at such speed that its movement would synchronize with the rotation of our planet, so that from the earth it would appear to stand still and hover always in the same position. This would make it continuously accessible for the transmission of audiovisual signals and at the same time more reliable for such work.

Based on this concept, Hughes-funded prototype satellite construction and tests led to a contract from NASA to produce and launch a satellite system. Accordingly, a satellite designed for synchronous-orbit communication, and hence called "Syncom," was hurled into space on July 26, 1963, and was eminently successful in taking up a position some 23,000 miles above the equator over the Atlantic Ocean, where its orbital speed kept it synchronized with the earth's rotation. At this

* A. H. Waynick, *ONRL Scientific Notes*, pp. 242–245 (July 31, 1974).

tremendous altitude, the orbiting time was not 2 hours but 24 hours, and because the earth itself rotates on its axis once a day—that is, *also* once every 24 hours—the satellite appeared to an observer on earth to be stationary. Because of this, it is referred to as a "geostationary" satellite (geographically stationary). The obvious advantage of this arrangement is that the ground stations can remain in continuous radio contact with the satellite without having to alter the pointing direction of their antenna beam patterns. Thus, if the satellite is "parked" over the Atlantic, the American ground stations and the European ground stations can be *permanently* aimed at the satellite, and the cumbersome variable-pointing requirement of the ground station for the more rapidly orbiting satellites is thus eliminated.

One disadvantage of the geostationary satellite is that the rocket required for elevating it to 23,000 miles must be much more powerful than the rocket for the 100-mile-altitude satellite. But the space

Fig. 126. A U.S. spaceship orbiting the moon (foreground) photographs the earth (upper left center) partially illuminated by the sun. Courtesy Communications Satellite Corporation.

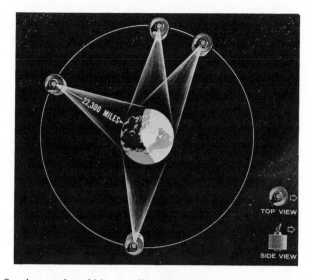

Fig. 127.   Synchronously orbiting satellites (stationary with respect to the earth) are now "parked" over the Atlantic, Pacific, and Indian oceans, providing worldwide television coverage. The Intelsat III model is sketched here.

Fig. 128.   A closeup of the Intelsat III communications satellite.

program had furnished such powerful rockets, and the success of Syncom pointed toward a bright future for communications satellites. Soon the U.S. satellite corporation "ComSat" (for Communications Satellite Corporation) was formed. Figure 126 is a cover photo of the 1969 Annual Report of that corporation, obviously taken from the moon. Soon ComSat was employing much larger rockets to put much larger communication satellites into geostationary orbits. By 1970, such satellites were positioned all around the globe (Fig. 127), providing worldwide television coverage and a sizable increase in the number of worldwide communications circuits (phone circuits). Figure 128 is a photo of Intelsat III (four of these are shown in the sketch of Fig. 127), which has a capacity of 1200 telephone circuits. Figure 129 shows a more recent and much larger communications satellite, Intelsat IV,

Fig. 129. The Intelsat IV is larger and provides a significant increase in transmission capacity over Intelsat III.

Fig. 130. This rocket and its boosters, capable of placing in synchronous (geo-stationary) orbit a (satellite) payload four times heavier than earlier boosters could, is used for launching the Intelsat IV satellite. Courtesy Communications Satellite Corporation.

which provides several times more circuits than Intelsat III. Figure 130 shows the Atlas Centaur rocket which was used to launch Intelsat IV; it has more than four times the payload capability of the earlier Delta launching rockets.

Four Intelsat IV's, which were launched in 1971 and 1972, have capabilities in excess of those of all the other satellites in the world, so that most communities in the world, including remote and inaccessible places which never had adequate communication facilities, can be in the satellite network. Intelsat IV's have had interesting tasks, including the transmission of color TV coverage of President Nixon's visit to China (February 1972) and of the Olympic Games.

DOMESTIC SATELLITES

In view of the significant benefits which communications satellites have given to nations all over the world, it is surprising that satellites to increase communications capabilities between cities *within* the United States took so long to come into being. Waynick has noted that even though an initial application to the U.S. Federal Communications Commission for a domestic system was made as early as 1965, "the U.S. Federal Communications is apparently following its usual practice of protecting vested interests,"* and nothing resulted. Appendix 61, taken from the 1969 Annual Report of the Communications Satellite Corporation, has as its caption: "The domestic service proposed by COMSAT would include high-capacity satellites in synchronous orbit capable of serving any known requirements and a network of earth stations as required in the United States." Five years after *that* proposal, there had still been no approval by the FCC.

Canada, however, did not need to bow to the U.S. FCC delays, and in 1972 Canada's first domestic satellite Anik I (*Anik* is Eskimo for "brother") was launched into a geostationary orbit. It links Canada's vast land mass, stretching across six time zones and north to isolated areas around the Arctic Ocean. Stationed at 114 degrees west longitude, the 1200-pound satellite and the earth complete a full revolution every 24 hours. The satellite is the first of three Aniks ordered under a $31 million contract by Telesat Canada, a special public and private corporation created in 1969 by the Canadian Parliament.

Anik has the capacity to provide more than 5000 two-way telephone circuits or 12 color television channels. It was developed and built by Hughes Aircraft Company, Segundo, California, and two Canadian subcontractors. Its signals are received by a ground station system comprising 37 Canadian-built stations, and a second, Anik II, was launched in 1973. In August 1974, RCA announced the use of Anik II to provide a U.S. domestic circuit, thereby initiating the nation's first domestic commercial satellite service connecting the East and West coasts. RCA used ground stations near New York and San Francisco. The *Wall Street Journal* article describing this development noted that "because a satellite connection can be completed with fewer

* *Ibid.*

relay points than can an earthbound system, the quality of transmission is better." A March 1974 article pointed out that "although the Soviet Union and Canada have both launched their own domestic satellite networks the U.S. had yet to do so on a civilian basis."* It described the above-mentioned RCA use of Anik II, and its title for a sketch of the coverage (Appendix 62) was "High-Time That U.S. is Serviced by Domestic Satellites."

By 1974 the parade toward the domestic satellite communications business had included Western Union, American Satellite Corporation, Communications Satellite Corporation, RCA, and IBM. In September 1974, the nation's first truly domestic communications satellite, Westar I, owned by Western Union, became operational. In 1975, the *Wall Street Journal* announced that it would use Westar I to operate, by satellite, the first newspaper production plant, located in Orlando, Florida. Facsimiles of full-size *Wall Street Journal* newspaper pages are now transmitted daily by satellite from another printing plant 1200 miles away.

In late 1975, the FCC granted a large number of approvals for domestic satellites, three to RCA† and one to a triparty venture, IBM ComSat General, and Aetna Life,‡ and it looked favorably on a joint system of AT & T and General Telephone.§ RCA's Satcom A was launched in December 12, 1975,‖ for a system to provide television, voice, and high-speed data transmission for the United States, including Alaska and Hawaii. To connect satellite to satellite, the U.S. Air Force has developed** a laser-relay satellite (Fig. 131).

DIRECT BROADCAST SATELLITE

One intriguing possibility for a communications satellite is that of relaying a television signal *directly* to the millions of homes whose rooftops would be under this manmade "star" continually hovering over them. Because the satellite is 23,000 miles high, there is no mountain

---

* "Domestic satellites starting to fly," *Industrial Research*, p. 30 (March 1974).
† *Electronic News*, p. 2 (November 10, 1975).
‡ *Microwaves*, p. 19 (November 1975).
§ *Wall Street Journal*, p. 4 (December 15, 1975).
‖ *Science News*, p. 392 (December 27, 1975).
** *Laser Focus*, p. 22 (January 1976).

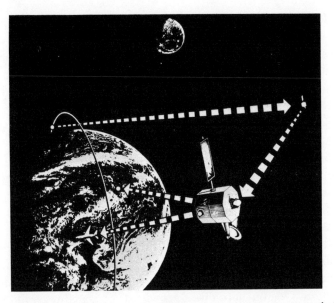

Fig. 131.   A laser-relay satellite capable of transmitting data at an extremely high rate. Three such satellites containing neodymium-YAG lasers will be placed in geostationary orbits 22,000 miles above the earth's surface during a 5-year period.

range high enough or valley deep enough to cause its signal to be blocked to any home in the entire United States. Such a satellite has received the name "direct broadcast satellite" because, unlike present satellites, it would broadcast directly to many homes.

It is interesting to compare the way such a satellite would furnish TV to the home with the way television programs reach the viewer today. Consider a program originating in New York City. It is routed over the AT&T networks (usually via many microwave relay stations) to hundreds of cities. In all of these many cities, separate individual TV stations then broadcast it to the listeners' homes. However, in the mountainous areas, homes in valley locations experience very poor reception because there are often hills located between them and the television broadcasting station. Accordingly, in such areas, "Community Antenna TV" systems (abbreviated CATV) have sprung up whereby one powerful receiving antenna receives the broadcasted signal and routes it by cable to the many homes desiring the service. Thus, to transmit one television program to millions of homes, one originating station is used, thousands of miles of AT&T networks are used, many hundreds of individual broadcasting stations are used, and many tens of CATV systems with many hundreds of miles of *their* cables are used. For the direct broadcast satellite, on the other hand, *one* originating station and *one* satellite are all that it needed. *All* homes in the urban, suburban, rural, mountain, valley, etc., areas are provided with excellent reception through the signal transmitted directly from the one satellite. The inherent line-of-sight path from the satellite to every home will assure that the quality of the television reception is superior to that now experienced in most U.S. homes today.

Thus when one adds up the cost of investment in TV stations, TV networks, and CATV hookups which such a satellite system would substitute for, the question of return on investment of such satellites becomes academic. Two and one-half million subscriptions, at $5 or more a month, for CATV privileges demonstrate that TV reception in most suburban and rural areas is not good. The amount of money this represents, just by itself, is equivalent to a 15% return on an investment of $1 billion. One or more direct broadcast TV satellites could reradiate a number of channels of high-quality signals to every home in America, thereby giving the average U.S. viewer reception as good as or better than that he now obtains by the present combined TV stations, TV networks, and CATV networks.

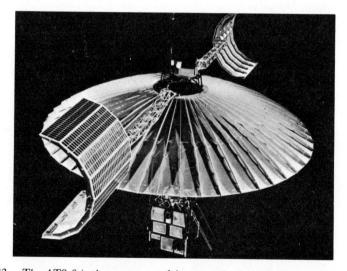

Fig. 132.   The ATS-6 is the most powerful communications satellite ever launched. Its 30-foot dish antenna focuses a beam of radio energy so intense that a small antenna can receive color television signals.

Fig. 133.   This low-cost satellite-receiving terminal can operate at C, S, and K bands and is in operation at 130 locations throughout the United States, receiving signals from ATS-6.

In May 1974, NASA launched Applications Technology Satellite-6 (Fig. 132), the most powerful, complex, and versatile communications spacecraft ever developed. Signals from ATS-6 are so powerful that they require only a modest-size TV receiving antenna (Fig. 133). Many companies predict that direct satellite-to-home transmission will be practical by 1979 at a receiver cost well within consumer means. Interest is therefore strong in an experimental 12,000 megacycle (12 gigahertz) broadcast satellite being built for Japan* which requires only a 4 1/2-foot receiving antenna at the home. It is estimated that a complete

* *Microwaves*, p. 22 (October 1974).

receiving package (antenna and converter) would cost between $300 and $700.*

How have these developments affected the way our earth appears at points in space that are quite distant from our solar system; in other words, what does planet earth look like from a distance because of the extensive and powerful radiation of microwave signals being sent to these many satellites? "A Million Times Brighter Than the Sun" was the title of a 1974 article in which it was noted that television stations radiate 200,000 times more radio energy than the earth itself, and radars are even *more* powerful, with the Arecibo (Puerto Rico) radar, if seen from deep space, appearing to be a million to a billion times brighter than the sun.†

In conclusion, let us note that a program having only the prescribed goal of landing a man on the moon by 1970 gave the world, in an unpredicted way, a class of satellites—the communications satellites—which portend a very large and very valuable expansion of the world's communications and television capabilities. A bright note on which to close this chapter is the launching, in mid-May 1976, of the first Bell System domestic communications satellite, for expanding telephone service in the 48 contiguous states of the United States and for taking over telephone traffic between the mainland and both Hawaii and Puerto Rico.

* *Ibid.*
† Frank D. Drake, *Industrial Research,* p. 49 (November 1974).

9

WHY INVENT?

Following the review, in the last three chapters, of innovative developments in various communications technologies, we discuss in this chapter the incentives involved, to assist those who might persuade others to *become* creative. In particular, we here review some of the possible rewards which could influence those who do become more innovative.

We consider first an incentive that should be attractive to most, largely because it can bring some measure of independence. Placing it first may seem to the reader to be stressing financial rewards too much, but, after all, freedom from want is a worthy goal for people and nations the world over. Another reason for placing this incentive first is that parents of future innovators may find this the most easily grasped reason for convincing youngsters that they should be creative.

WHEREWITHAL

The existence of patents and copyrights provides a means for protecting creative expressions, thereby ensuring innovators that their handiwork will be rewarded in *some* monetary way. Even in industry, where patents or inventions made by an employee are automatically assigned to the company, the creative engineer reaps the benefits of his inventions through better salary and promotions. Because his patent record is publicly available, a person's creative ability is known to many, and offers of high-salaried positions by other companies are therefore always available to the creative engineer. His own company is fully aware of this situation, and it therefore makes certain that the inventor's salary and the prestige of his position match those of the marketplace. (I recall keeping a file of close to a hundred offers made me by other corporations and by universities.)

If the inventor is fortunate enough to hit upon a valuable new idea while still a student (that is, before joining a company), he can then *own* (without assignment) patents associated with his idea. Under these circumstances, companies can be talked into giving royalties on products manufactured by them which use his ideas. Another arrangement is to enter into a contract whereby the patent or patents involved *are* assigned, but the company agrees to pay royalties. Since the time between the filing of a patent application and the issuance of the patent may be several years, the inventor often need only wait for the first patent office action to take place (if it shows a favorable reaction) to enter into negotiations with an interested company. The mutually agreed upon contract would then spell out the royalties to be paid *should* the patent issue and *should* the product be manufactured. (If manufacture does not materialize, the contract usually specifies that ownership of the patent rights reverts to the inventor.) Appendix 63 shows a royalty payment from the D. H. Baldwin Company in connection with my patents on the Baldwin electronic organ (this development was discussed in some detail in Chapter 3). Although about ten organ units had been built in the late 1930s, World War II prevented manufacture of them until after the war; this is why the 1947 date appears on the letter.

Even in the situation just discussed, the inventor often cannot rest on his laurels with one invention alone. The lifetime of a patent in the United States is 17 years, with no renewals allowed (a copyright, on the other hand, runs for a longer period and can be renewed). Since the task of preparing the new product for manufacture may take some years, the actual "remunerative" length can be much less than the 17-year patent life. Thus, in my Baldwin case, World War II reduced the 17-year protection by 7 years. Furthermore, it often takes the purchasing public some years to accept a new product, so that the sales start slowly, with those at the end of the 17-year period usually being the highest. I recall quite vividly the increase, month after month, in the size of my Baldwin royalty checks, until, boom, the 17-year period ended. As it happened, my income did not drop completely to zero, because there had been some foreign filings on the concepts, and these still had a year or two to run after the U.S. patent had expired. The much smaller royalties on them continued also to rise, but finally they too stopped, bringing the wherewithal aspect of my Baldwin association to an end. Meanwhile, however, my inventions had continued at Bell Laboratories, and my increasing salary there (as discussed above) quite effectively maintained my "freedom from want."

One aspect of royalty payments is worthy of brief mention. If an arrangement between company and inventor is made whereby a fixed (constant) royalty amount is paid per item to the inventor regardless of the selling price of the product, the inventor's royalty income from this contract is classed as a long-term capital gain (rather than earned income). In this situation, the inventor pays less income tax on that royalty portion of his income. My contract with Baldwin was made on this basis, but the signing occurred before the court had ruled that such fixed-amount-per-item royalties were to be considered as long-term capital gain items. When the ruling did take place, I approached the Internal Revenue Service, asking for an income tax refund based on the ruling. I was politely informed that the IRS did not recognize this (lower) court's ruling but that I was fully entitled to appeal this IRS nonrecognition of the court's capital gains decision. (The IRS examiner pointed out that he was aware of the author's Bell Laboratories salary, and he therefore felt that I could hardly afford the legal expenses involved in such an appeal!) Shortly thereafter, however, the IRS did accept the court's ruling, and full payment of my entitled refund (based on my earlier classification of the royalties as earned income) was made.

As a final reference to the affluence which an inventor can acquire, we need only recall the class in society reached by Thomas Edison, Alexander Graham Bell, Charles Kettering, and other famous inventors. My parents were in a low to medium income bracket, and so I was pleased to be able to afford the home in Ann Arbor, Michigan, shown in Appendix 64.

RENOWN

Most innovators also enjoy fame as well as fortune. In industry, this acquiring of renown comes about fairly automatically, because the company for which the creative engineer is working benefits substantially, in numerous ways, through its publicizing of the new concepts and achievements of their inventors. The company's customers look more favorably on products of an innovative corporation, and such publicity also induces other invention-conscious engineers to join the company in one of its research operations.

Figure 136 shows one of the results (in the magazine *Newsweek*) of a Bell Laboratories press release; as it was seen by many of my friends and associates, it did provide a form of incentive for dreaming up more new ideas. Appendix 65 shows that the Bell story was picked up and used by publications in many parts of the world. We saw in Appendix Item 47 that the photo of Fig. 134 was used on the cover of one magazine; its story, like the one of Fig. 134, included the names of F. K. Harvey and W. E. Kock.

Because innovative persons often are given various awards, the company which employs them also provides them with prestige incentives by issuing press releases describing such awards. Appendix 66

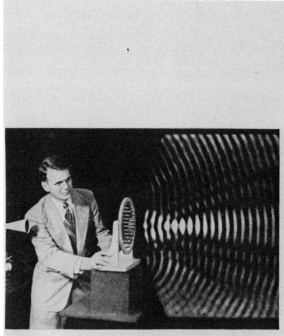

Seeing Sound

One Saturday morning early last April inspiration came to Winston E. Kock. At home for the week end in his remodeled Colonial farmhouse in Basking Ridge, N. J., the Bell Telephone Laboratories acoustical engineer thought of a device to make sound waves visible.

Would it work? Kock set about putting his idea into mechanical form. Fazed not at all by lack of laboratory facilities, he borrowed an Erector set from Winston Jr., aged 10, and constructed a crude but convincing model. The following Monday Kock took his contraption to Bell Labs at Murray Hill, N. J. There he and an associate, F. K. Harvey, developed a full-scale machine.

Last week they put the now perfected gadget through its paces and revealed how it works. Mechanically, it is simple. Small motors sweep a metal arm up and down through a 3- or 4-foot arc. At the same time the arm advances horizontally. The net effect is to scan a plane, as a television tube scans its screen.

A sensitive hearing-aid microphone, about the size of a quarter, and a quarter-watt neon lamp, no bigger than a flashlight bulb, are mounted at the tip of the arm. Noise picked up by the microphone is converted to electric current and fed to the lamp. As the sound level varies, the brightness of the light varies in proportion.

The final effect as photographed with a time exposure, is a visible pattern of the sound waves through which the microphone passes (see cut). The picture shows Harvey and the sound pattern produced by a loudspeaker (left) and

Harvey and a pattern traced by a folded ribbon of light 1,000 feet long

Fig. 134.   The Newsweek article stemming from a Bell Laboratories press release.

shows a typical Bendix Corporation press release of this type. Appendix 67 shows an earlier award certificate, based on a development which we shall discuss in Chapter 10.

The wide coverage which often results from Corporation press releases benefits both the innovator and the corporation. For example, 38 publications picked up the story on the metal microwave lens of Fig. 100. Also, long after I had left the Baldwin Piano Company, that corporation felt that it would gain by having the formant concept associated with their electronic organ. As we noted in Chapter 3 (Appendix Item 9), the formant patent was selected by the American Patent Law Association for its 1976 U.S. Bicentennial program. Appendix 68 is the draft of that portion of the Patent Associations essay, showing that both I and the Baldwin Piano Company received recognition. Appendix 69 shows a Bendix Corporation advertisement which appeared in the *Wall Street Journal* and which described some of the antisubmarine research conducted by my Bendix group.

Even in the early stages of creativity, publicity leading to repute can occur. We have discussed chess game tournaments and chess problem composing tournaments. The winning of chess games can often lead to the awarding of trophies (Appendix 70) and newspaper coverage (Appendix 71), and the publication of an outstanding chess problem can also lead to repute, particularly when the problem is unusually difficult. Appendix 72 and 73 are newspaper items which note the difficulty a large chess-solving group had in solving two chess problems.

HELPING OTHERS

Creative engineers are important assets, both for their employers and for their country. Accordingly, for their innovative contributions, they can enjoy the satisfaction which accompanies their service to the nation in improving the economy. We stated in the Preface that the economic growth and strength of a nation are directly related to the ability of its people to make discoveries along with their ability to transform these discoveries into useful products, and that the large increase in output per capita in the United States has been held to be attributable to technological advances (innovative improvements). Dr. David Ragone, Dean of Engineering at the University of Michigan, has noted that "Historically, the strength of the United States has been as innovator . . . Kettering is among the greats, and Edison."

In a 1955 interview by Donald Robinson of Bell Laboratories President Mervin J. Kelly (entitled "Should Your Child Be an Electric Engineer"), Dr. Kelly noted that "the rewards of the profession are large" and that, over and beyond the good salaries,

> the average electronic engineer gets a lot more out of his profession than his salary. I know I do. There is a challenge in electronic engineering that is hard to match—the challenge of taking a nebulous idea and making it assume shape as a constructive, operating reality. I can remember as vividly as if it were today the thrill that came when a group of us Bell engineers were working on the first trans-Atlantic radio telephone. The problem was a tough one. The biggest vacuum tube available could produce only 250 watts. We had to develop a tube capable of twenty to twenty-five times that output. It was my good fortune to be able to design such a tube. Then we had to devise a system that would use this tube to transmit voices unfailingly across the 3,000 miles of static-ridden Atlantic water. Finally, the word came back. Our messages were getting through. Trans-Atlantic telephone service was assured. I was a very happy man.
>
> There are still other rewards. The electronic engineer has the respect of his community as a well-educated professional man. He has the gratification of doing work which is beneficial to mankind. Just think of the lives that have been saved, for example, by the X-ray tube. In these days of international stress, he also has the satisfaction of knowing that he can make a marked contribution to the defense of democracy. More than one-half of the electronic industry is now engaged in defense work. The continental defense system is all electronic. The brain which makes the Army's land-to-air missile Nike so effective in hunting out enemy planes and destroying them is electronic. The instrumentation of atomic weapons like the H-bomb, the A-bomb and the warheads on missiles is electronic. In fact, the electronic engineer is as essential to atomic weaponry as the nuclear physicist.
>
> Actually, an engineering education starts in high school. A boy who wants to be an engineer should study all the mathematics and physics he can get there.
>
> I am often asked, "Can girls make a go of engineering?" The answer is a resounding "Yes." There are some very good women engineers who hold down splendid jobs and do fine, creative work. This is particularly noticeable in electronics.
>
> To the girls as well as the boys who have the requisite aptitudes, I say: by all means, select electronic engineering for your career. You can help create a better, more livable world."

The importance to the United States of technological advances is well described in a publication of Gould, Inc.:

> Technology turned the United States from a wilderness which few suspected could be held together into one of the greatest nations on earth.

Technology can continue to keep the United States in that happy position. To turn away from technology is to lose that which has been the best hope of mankind for two centuries. In whichever direction we look, we see plainly and unmistakeably, that American strength, American comfort, and American liberties can be maintained and advanced only through a studied and wise continuing advance in technology.

Bell Laboratories President M. J. Kelly was honored in 1958 with the award of the National Security Industrial Associations' James Forrestal Medal, and in his address he noted that

Research and development have been the driving force in accelerating an expanding economy. The rewards to our society such as higher living standards, better health and increased span of life are familiar to all.

and stated further that

I stress the need for our secondary schools to identify a larger fraction of our youth, specially endowed for careers in science and technology, and to interest them in the career.

As a closing example of how engineers help others, we should mention television. The way that television has entered the lives of everyone must surely give the many creative engineers involved in its development a warm feeling of satisfaction. One is reminded of the words in the Prayer of Thanksgiving suggesting that we show our thanks "by giving up our selves to Thy service." And "serving mankind" could be one interpretation of this last phrase, a service which innovators of technology provide.

EMINENCE

We now review another incentive for being creative, the fact that others generally recognize the innovative engineer as being one who could be of value in other endeavors. His reputation thus leads him to be selected for various appointments, and he soon becomes regarded as an "eminent" man.

One such type of recognition is an appointment to a board of directors of a corporation (Appendix 74). Such appointments benefit both the corporation and the innovator, the corporation benefiting from the appointee's ideas and his contacts and the innovator learning of new developments which he might otherwise miss. Appointment to boards of trustees of colleges and universities also is made (Appendix 75), with similar benefits to both parties as mentioned above. A church will often find it valuable to appoint innovators to its board of elders. I had the valuable experience of being so ordained when Robert McNamara was also a Ruling Elder and the President of Ford Motor Company (Appendix 76). McNamara later became Secretary of Defense; he is presently the President of the World Bank.

Universities regularly bestow honorary degrees on their graduates who have distinguished themselves in one way or another. Here too both the university and the recipient benefit, the university from the publicity which usually mentions the accomplishments of its own graduate and the recipient from the new contacts he makes at the award ceremony and from the publicity he receives, which alerts his employer to the university's recognition of his talents (Appendix 77).

As the inventor continues to contribute, awards to him also continue. The one shown in Appendix 78 was made after I had left Bell and joined the Bendix Corporation; it was given by the honorary electrical engineering fraternity, Eta Kappa Nu. Appendix 79 indicates my election to a special class of membership and the small insert describes what the editor of the monthly journal of the fraternity, the *Bridge,* calls a "grand slam," the first in the fraternity's history.

The last item in this eminence section is a mention of my election as a Fellow of the Indian Academy of Sciences. This was a most welcome election because in 1973 there were only 51 such Fellows, with 29 of these having won the Nobel Prize (Appendix 80).

THE FRATERNITY OF DOERS

A strong incentive that should be stressed when parents seek to convince their children that they should try to become creative is their later ability thereby to "join the club." We refer to a completely unchartered fraternity of persons who, largely through their creativity, have been recognized by the other "members" as "doers," as people of accomplishment. It is natural for such a fraternity to exist, because creative people have a common bond with others who have been successful in reaching this same goal.

Since the doer often is recognized when still fairly young, his joining the club can occur fairly early. Appendix 81 lists, for example, the persons attending the Institute for Advanced Study during 1935 and

1936, and it is fairly obvious why I felt most fortunate in having been accepted (probably because of my earlier studies and research in Berlin). It is seen that the list includes Einstein, Von Laue (my examiner in Berlin), Von Neumann, Pauli, Weyl, Veblen, and many younger members who have since become renowned, such as Taub, Pryce, Hoffmann (recipient of the American Institute of Physics Award for his famous book on Einstein*), Breit (we discuss his early radar contributions in the next chapter), Levinson, Rosen, and many others.

The election to Eminent Membership in the Eta Kappa Nu fraternity, discussed above, placed me among famous electrical engineers who have been similarly chosen. These included Lee DeForest (the vacuum tube), Vladimir Zworykin (television), John Bardeen (the transistor), M. J. Kelly (President of Bell Laboratories), Lee Dubridge (Director of the MIT Radiation Laboratory), J. B. Wiesner (now President of MIT), and numerous others.

In 1950, I was asked to participate in a highly classified Summer Study named "Project Hartwell." It was supported by the U.S. Navy, and its Director was Professor Gerrold Zacharias of MIT. Participants included, among others, Merle Tuve (who, with Gregory Breit, conducted very early radar experiments as discussed in the next chapter), Luis Alvarez (we discuss his Vixen innovation in the next chapter), J. B. Wiesner, Lloyd Berkner, Ivan Getting (President of the Aerospace Corporation), E. E. David, Jr., and my first and second bosses at Bell Laboratories, H. T. Friis and R. K. Potter. I dedicated one of my books to each of these last two men.† I also dedicated one to the Hartwell Director J. R. Zacharias, to show my gratitude for having been selected as a member of the "Hartwell Club."‡ Appendix 82 shows a letter from President Truman expressing his "deep appreciation of the [Hartwell] Group's splendid accomplishment."

Participation in committee work by members of the doers fraternity usually leads to new ideas. Recently I was asked to be a member of the

---

* Banesh Hoffmann, *Albert Einstein—Creator and Rebel* (New York: Viking Press, 1972).

† Winston E. Kock, *Sound Waves and Light Waves* (New York: Doubleday Science Study Series No. S40, 1965) (translated into German, Russian, Italian, Japanese). Winston E. Kock, *Seeing Sound* (New York: Wiley, 1971) (translated into German, Russian).

‡ Winston E. Kock, *Radar, Sonar, and Holography* (New York: Academic Press, 1973).

Visiting Committee of the Physical Sciences Division of the University of Chicago, which includes, among others, Wernher von Braun (Appendix 83), E. E. David, Jr. (earlier the President's Science Advisor, now President of Exxon Research and Engineering), J. B. Fisk (retired President of Bell Laboratories), and Nobel Laureate Luis Alvarez. The Dean of the Physical Sciences Division and host to the Committee, Professor Albert V. Crewe, recently was given wide TV and news coverage about the first movies which he made of atoms *moving* in their natural state. In 1971, Crewe had been selected by the Journal *Industrial Research* as "Man of the Year in Research" for his development of the field emission scanning microscope. At a 1975 meeting of the Committee, Luis Alvarez, at lunch, showed those of us seated near him a piece of *circularly* polarized polaroid material and what it could do. I later dreamed up a use for this material based on some earlier radar experiments. This concept will be discussed in the next chapter; it is mentioned here to show one of the benefits of meetings involving members of the "contributors' club."

Members of the club value opportunities to meet with others, and accordingly invitations to participate in special meetings are almost invariably accepted. Appendix 84 is a photo of three Nobel Laureates taken at my Ann Arbor home on the evening of a small symposium held at the Bendix Corporation's Research Laboratories in Michigan.

Members of the club often correspond with one another, particularly when one member has come up with an interesting idea. Appendix 85 shows such a letter, written by Bell's Lloyd Espenschied, inventor of the coaxial cable. In the letter he mentions the microwave waveguide innovations of Bell's George Southworth. I prize that letter and many others from my colleagues.

The possession of interdisciplinary ability often results in meetings between creative people in different fields. The outstanding violin artist Zino Francescatti visited Bell Laboratories, and, at my suggestion, agreed to play a portion of a violin solo on two different violins, one a valuable Guersan and the other an inexpensive one. The experiment took place (Appendix 86) in the Bell Laboratories "free-space" room (Fig. 120), and the recording of the two solos permitted a later analysis leading to some visible differences in the two being unearthed.*

---

* Winston E. Kock, *Seeing Sound* (New York: Wiley, 1971) (translated into German, Russian).

Correspondence often follows meetings, and the receipt of such letters from innovators from other countries is always a pleasant experience. Scientific meetings with foreign innovators also often lead to social meetings. Appendix 87 is a photo taken at the residence of U.S. scientist John Ide in Washington, following 4 days of meetings of delegations from the United States and Japan, who participated in a National Science Foundation sponsored United States–Japan Seminar, "*Information Processing by Holography.*"* *Applications of Holography* We shall discuss this subject in Chapter 11.

We close with two amusing items. Appendix 88 is a copy of an envelope that I received; the opposite, flap side was properly sealed with two blobs of sealing wax, each stamped with "The Seal of His Majesty's Ships." Appendix 89 shows the celebration of a motion picture premiere held in Boston when I (holding a letter with the producer) was the Director of the National Aeronautics and Space Administration's Research Center in Cambridge, Massachusetts.

Having reviewed some of the reasons why inventing leads to various kinds of rewards, let us return to a discussion of three more innovations and review the circumstances which led to their coming into being.

---

* *Applications of Holography,* ed. by Barrekette, Kock, Ose, Tsujiuchi, and Stroke (New York: Plenum Press, 1971).

10

RADAR

EARLY HISTORY

An extremely significant innovative development occurred just prior to World War II. It received the name *radar* from the letters *r*adio *d*etection *a*nd *r*anging. Now the transmission and reflection properties of radio waves had been conclusively demonstrated experimentally by the creative German scientist Heinrich Hertz as early as 1887. Following the invention of the three-electrode vacuum tube by U.S. engineer Lee DeForest, a mushrooming of radio technology occurred, with such developments including radio communication, radio broadcasting, and, later, television broadcasting. The field of radar, utilizing the *reflection* of radio waves, has now become comparable in importance and breadth of applications to radio communication and radio broadcasting.

Thus radar, originally developed as a means for providing the military with information on the location and movements of enemy forces, has since found wide application—thanks to numerous innovators in many fields—in civilian sectors, including sea transport, air traffic control (at airports), and aircraft navigation. This last use includes weather radars, with which aircraft pilots can detect and thereby avoid atmospheric thunderstorm regions (usually having high turbulence which would cause a shaking up of the passengers). Even automobile traffic is beginning to benefit from various applications of radar.

THE BRITISH BEGINNINGS

It appears that the earliest and most significant developments in radar occurred both in Great Britain and in the United States. As early as 1935, the British scientist Robert Watson-Watt submitted to the British Air Ministry a very important document on radar. (For the extremely significant part which he played in the initial and later radar developments, he was knighted by the British government in 1942, thereby becoming *Sir* Robert.) His document was entitled "Detection of Aircraft by Radio Methods," and the Air Ministry, recognizing its importance in military applications, immediately classified the document "Secret." (The first Secrecy Order which I received was placed on a radar patent application filed in October 1941. See Appendix 90.) Watson-Watt's proposal was successfully demonstrated during some Royal Air Force maneuvers in 1937. Many years after the war, it was reported that Sir Robert, while driving through Canada, was stopped because a *police radar* showed that he was speeding!

Following the outbreak of World War II, British scientists hit upon the design of an electronic vacuum tube which could generate very high output powers for the short-wavelength radio waves called "microwaves." U.S. radar designers welcomed this innovation and promptly put the new tube, called a "magnetron," to good use in U.S. radar.

THE U.S. BEGINNINGS

Gregory Breit and Merle Tuve of the Carnegie Institution determined, in 1925, the height of the ionosphere, using short pulses of radio waves. During the 1930s, the U.S. military and, as we noted earlier, the Bell Telephone Laboratories made great strides in advancing the microwave art (such as metallic waveguides). From various Bell publications, the following account of the early U.S. radar activity was put together. From 1934 through 1937, radio detection and ranging— i.e., "radar"—underwent development at the U.S. Naval Research Laboratory (1934) and at the U.S. Army Signal Corps Laboratory (1935). This early radar work employed transmitters and receivers using large antennas (over 20 feet) and long-wavelength radio waves. It was clear that much higher frequencies would be required to extend the potential of this new technology from early-warning applications to the accuracies needed for control of gunfire. Higher frequencies would permit narrow antenna beamwidths and manageable antenna structures aboard ship. Yet components for higher frequencies were not then available.

With this need in mind, the Navy approached AT&T in 1937 requesting assistance from the Bell Laboratories on a secret project of great importance to the nation. (The Navy was aware of the pioneering work in radio research at the Holmdel and Deal, New Jersey, laboratories of Bell in pushing the usable frequency spectrum upward for communications systems and of the associated research on directive antennas.) The other extremely significant contributing organization to World War II radar, the MIT Radiation Laboratory, was organized in October 1940.

Because of the obvious importance to national defense, AT&T authorized Bell Laboratories to explore possibilities and to report to the Navy on the results of the investigation. Accordingly, in early 1938, a small group of highly skilled engineers was assigned to Bell Laboratories' Whippany, New Jersey, laboratory, about 30 miles west of New York City. The isolated rural setting met requirements for absolute secrecy and had the open terrain needed for field testing. The group had to design and build their own instruments; none was commercially available in the frequency range to be explored. Transmitters, receivers, antennas, and new cathode-ray indicators were devised, and a model was built to show what might be done.

In July 1939, the crude model, operating at 700 megahertz, was set up at a site on an 80-foot bluff at Atlantic Highlands, New Jersey, overlooking New York Harbor (Fig. 135). Late into the night following the installation, Bell Laboratories engineers were thrilled to plot quite accurately the movement of ships coming and going in the Ambrose Channel—not fully knowing the impact such tracking would soon have in the naval battles in the Pacific. Later in the month, the equipment was successfully demonstrated to the Navy and Army and to high officials of the Bell System. The demonstration resulted in a Navy contract with Western Electric for a production version to be developed by Bell Laboratories. The first of these, known as the "Mark I," was installed on the cruiser *Wichita* in June 1941, and several more installations were under way on ships of the fleet at Pearl Harbor when the Japanese attacked on December 7, 1941.

Radar was a major contributor to victory in World War II. About half of the total Bell Laboratories military effort was devoted to developing about 100 major radar types during the war. Included were all the fire control radars on Navy ships, all but one of the five submarine radars, mobile radars for the Marine Corps, various types of navigation

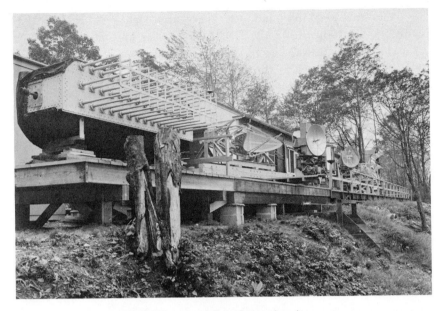

Fig. 135.   An early radar-testing site.

and bombing radars for the Air Force, and Army radars including an integrated search and, for the first time, fully automatic track radar.

Radar also proved extremely valuable in ship-versus-ship encounters. In the autumn of 1942, the *U.S.S. Boise* sank six opposing war ships in a night battle through its use of radar. The battle took place on October 11 off Savo Island near Guadalcanal. Even in the intense darkness, the *Boise* was able to contact the enemy with its surface search radar equipment, an early type developed at the Bell Telephone Laboratories and manufactured by the Western Electric Company. As explained by Lieutenant Commander Philip C. Kelsey, "We contacted the ships at 13,500 yards, according to our surface search radar, and thus provided the fire control party with initial target bearings. The enemy ships, apparently unknowing in the darkness, moved closer and closer to the *Boise* until a scant 3900 yards separated us." The *Boise* opened fire at almost point blank range, and in 27 short minutes the engagement was over with none of the enemy ships remaining afloat.

The Bell Laboratories test facility at Atlantic Highlands, New Jersey, overlooking Sandy Hook Bay and the entire New York Harbor (Fig. 135), played a great part in the convincing of high-ranking Naval officers, particularly those to whom radar was an unknown quantity in the early days of the war, of the fact that radar could accurately direct the pointing of ships' guns. The actual demonstration of fire control radar was found to be a much more satisfactory method of instruction than words, charts, and pictures could be. At the left of Fig. 135 is one of the first Naval fire control radars (see Fig. 74).

RADAR FUNDAMENTALS

In the most common form of radars and sonars, short pulses of energy are radiated periodically, usually from a highly directional radiator, and immediately after each pulse is transmitted the radiator is made to act as a receiver. If a reflecting object is located in the beam of the radiator, the energy in the pulse will strike it and will be scattered in many directions. Some of the energy, however, will be reflected back toward the radiator (now acting as a receiver) as a greatly weakened but still short pulse of energy. This received pulse is amplified and displayed to the radar or sonar operator.

Figure 136 portrays four different instants occurring in a water wave (ripple tank) arrangement that simulates the action of a radar. In (a), a pulse of energy is generated corresponding to the radar pulse and producing the waves shown. In (b), the waves are starting to move toward a model of an aircraft. In (c), the waves have reached the target and some reflection has started to take place. Finally, in (d), the reflected energy has returned to the vicinity of the transmitter, which now could be acting as a receiver, thus notifying an operator of the presence of a target. (This photo is obviously not to scale.)

Because the velocity of propagation of radio waves and sound waves is known, the distance or range of the reflecting object can be determined from the length of time taken by the echoed pulse in returning to the receiver. Thus for radio waves, which travel at the speed of light (186,000 miles per second), a reflecting object at a 9.3-mile range (corresponding to an 18.6-mile round trip path) would generate a returned pulse at the receiver 1/10,000 second after the transmitted pulse was radiated. Objects at other ranges generate received pulses at different times, so that the ranges of all objects which lie within the transmit–receive beam can be determined.

If now the direction in which the beam is pointed is altered slightly, other objects, lying in the new beam direction, will generate

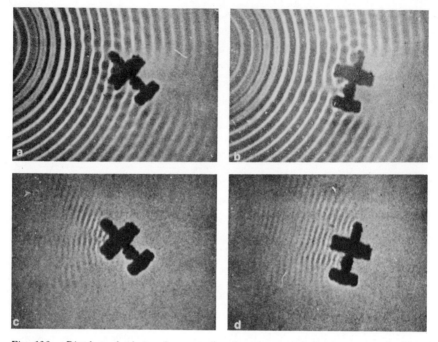

Fig. 136. Ripple tank photos show a pulse of wave energy being sent out and reflected back by a model. Courtesy U.S. Naval Research Laboratory.

"echoes," and the range of these can be also made available to the operator.

THE PLAN POSITION DISPLAY

Figure 137 is a sketch of the display available in a typical marine radar. It represents the face of an electronic device called a "cathode-ray tube." In this device, which is very similar to the well-known "picture tube" of a television set, a very narrow, movable beam of electrons is directed toward the rear face of the circular glass plate of the figure and this face is coated with an electrosensitive phosphor that glows when struck by the electron beam. The vertical line in the figure corresponds to one sweep of the electron beam, as it proceeds from the center of the circle upward toward the edge. The strength of the electron beam, and therefore the brightness of the visual signal it generates on the phosphor, is controlled by the strength of the echoes as received and amplified. Thus, along the vertical line shown, only one echoing area exists, the island labeled A. The speed of motion of the electron beam is made to match the range coverage desired. Thus, if the maximum range desired is, say, 20 miles (so that island A might be at 12-mile range from the ship carrying this marine radar), the speed of the electron beam is made to match the echo travel time corresponding to an echo at 20-mile range.

The beam is now pointed in a new direction, say, slightly to the right of the straight line, and a new pulse is transmitted. Another set of echoes results; again, in this case, it is only from island A. As the beam continues to move clockwise, it eventually covers the entire circle and starts over again. This form of display is called a "plan position indicator." The phosphor in the cathode-ray tube of such a radar display is made to have some "persistence" so that the entire picture dies away only gradually. In the sketch a second island B is indicated, and a smaller object (really a dot) corresponds to an aircraft. For this last echo, a sizable movement of this dot will be noticeable in each new complete radar picture, whereas the island will have a motion (relative motion) corresponding to the slower speed of the ship.

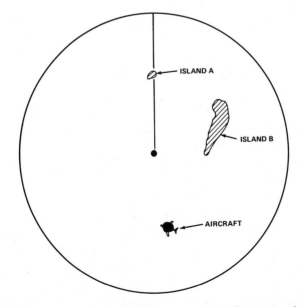

Fig. 137. A plan position indicator display as employed on a marine (shipboard) radar. The ship carrying the radar is located at the central dot.

It is evident that the detail which can be incorporated in the displayed "map" will depend on (1) the sharpness of the transmit-receive beam (this determines the angular or azimuth resolution) and (2) the pulse length (this determines the range resolution).

THE A-SCOPE DISPLAY

Historically, the oldest radar display is the A-scope display. It indicates target range and, within limits, the amplitude of the target echo. The A-scope, which is simply an oscillographic presentation of the radar return, is used when angular information is not required or is provided from another sensor. For example, the A-scope might show range to targets directly in front of a radar-equipped aircraft or to targets whose angular position is determined for the radar with an optical sight or infrared sensor.

Figure 138 portrays an A-scope presentation of one of the early milestones in radar history. Occurring in January 1946, the event marked the first time a radar echo was received from the moon. As the upper sketch shows, the transmitted waves traveled 238,857 miles on their way to the moon and an equal distance after they were reflected. Traveling at the speed of light, they took only 2½ seconds to cover their long path.

At the left of the A-scope record in the lower part of the figure, the writing beam is still being influenced by the very powerful outgoing pulse. After the pulse is on its way, however, (indicated in the upper sketch by the sets of four lines in the transmitting beam), the writing record shows, through its nearness to the lower scale of miles, that no echoing objects are in the beam. The echo from the moon is the broad plateau at the right. In between the outgoing and echo pulses, the writing beam record is quite irregular. This is caused by noise, coming possibly from outer space, but also from circuits within the radar itself. This moon experiment was under the direction of the U.S. engineer Lt. Col. John H. Dewitt, Jr. The newspaper article describing the experiment also hinted at "radar spaceships," which are now a reality.

Another form of radar or sonar display is the one described above in connection with the marine radar. Called a "plan position indicator" (PPI), it displays bearings to the targets as angles and ranges as the radial distance to the target echoes. For ground installations, the PPI angular reference is fixed and usually north oriented. Airborne and

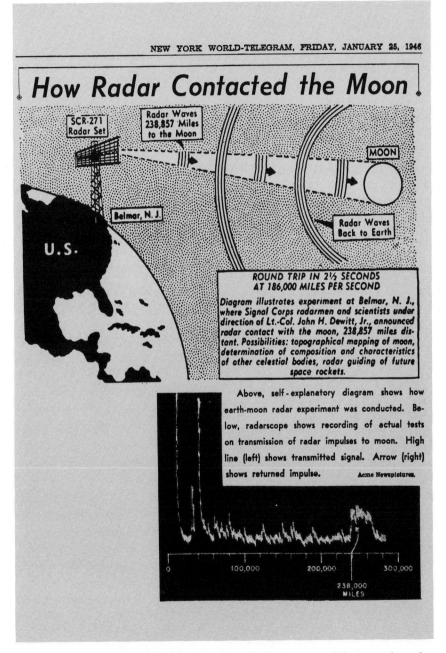

Fig. 138. A newspaper clipping announcing the first contact of the moon by radar. The A-scope presentation is shown below.

shipborne PPI displays can be either north oriented or vehicle oriented, depending on the application.

FIRE CONTROL RADARS

In Figure 139, an 8-foot experimental radar paraboloid is shown. The photo was taken on May 25, 1943, when the first complete systems test of a fairly important World War II radar was being conducted at the Radar Research Laboratory of the Bell Telephone Laboratories at Holmdel, New Jersey (I am the central figure, looking toward the antenna), and Appendix 91 is a page from the patent issued on that radar antenna. This radar was an experimental model of the early Mark 13 fire control radar, destined to be used to direct the main batteries of

Fig. 139. An 8-foot parabolic reflector designed at the Bell Telephone Laboratories for a World War II shipboard radar. The person at the right, peering into the lower cathode-ray tube, is Dr. Harold Friis, who for his creative contributions later received high honors from his native country, Denmark.

Fig. 140. The experimental antenna in Fig. 139 as mounted for actual sea tests.

battleships and cruisers. Another view of this experimental version is shown in Fig. 140. The tubular structure is seen to have attached to it a horn feed, and in operation the entire 8-foot reflector and its feed horn were made to oscillate in the horizontal plane, with a scanning motion of ± 5.8 degrees at an oscillation frequency of 5 hertz. The choice of these parameters was dictated by the radar mission; it was to display a small sector (11.6 degrees wide in azimuth, as contrasted with the full 360 degrees of Fig. 137), but it was to display the information in that sector rapidly (five back-and-forth scans, or ten total scans, per second). This enabled the radar operator (and the fire control officer) to "see," in both range and azimuth, the plumes of water that arose as the main-gun battery shells struck the water. Since the radar also "saw" the enemy ship under attack, the fire control officer could then adjust his gun batteries so that the next salvo of shells would zero in on the ship under attack. All of this information was available during the black of night, during fog, etc. Because of its great effectiveness to the Navy, this radar was highly commended by Rear Admiral G. F. Hussey, Jr., then Chief of the Bureau of Ordnance (Appendix 92). In a letter dated June 25, 1945, he stated: "While fleet experience with the radar equipment Mark 13 is limited to only a few ships, reports from these ships indicate that it is considered the best radar equipment yet installed on ship-board." Figure 141 is a closeup of this production model.

Fig. 141. The production model, called the "Mark 13," of the experimental radar in Fig. 140.

OTHER FORMS OF MILITARY RADAR

During World War II, a submarine radar was developed that enabled a submarine to remain almost completely submerged and still employ the radar in the dark of night when conditions for visual observation by means of a periscope were not favorable. In radars and sonars, the ability to steer the beam electrically—that is, without mechanically moving the radiator—permits much larger arrays to be used. These arrays are called "phased arrays." The very large, several-stories-high radar of Fig. 142 certainly does not lend itself to beam steering by mechanical rotation of the structure. However, because it is equipped so that all of its thousands of radiating elements can be properly "phased"—that is, energized with proper amounts of delay—its beam can be steered quite effectively by electronic means. This radar, called the "AN/FPS-85 Spacetrack Radar" or the "Spadat radar" for short, was built by the Bendix Corporation for the U.S. Air Force. It is a high-power phased-array radar capable of automatically detecting, tracking, and cataloging hundreds of space objects simultaneously.

DOPPLER RADAR

We discussed the Doppler effect for sound waves in Chapter 3, also noting very briefly there, in connection with Fig. 38, the use of that effect for radars for automobiles. This use is possible because radio energy reflected from a target moving toward (or away from) the radar similarly exhibits an apparent change in frequency upon reception. A particularly valuable use of this phenomenon is found in the aircraft navigation technique called "Doppler radar" or "Doppler Navigation."

In Doppler navigation systems, four very narrow beams of microwave radar energy are radiated downward from the aircraft. The frequencies of the four reflected signals are continually compared with the frequency of the radiated signal, and the differences, due to the Doppler effect, are determined. If there is no crosswind (which would cause the

Fig. 142. A multistory radar shown under construction for the U.S. Air Force by the Bendix Corporation.

plane to shift sideways, even though heading in the desired direction), then the echoes in the two forward beams will experience equal up-Doppler and those in the two rear beams equal down-Doppler. When a crosswind exists, these are *not* equal. Similarly, if there is a headwind, the plane's air speed will be different from its ground speed. But this will be revealed to the pilot by the amount of Doppler generated in the echoes from the ground.

When a crosswind exists, if the pilot (unaware of this wind without such a radar) continually flies, say, due east, he will find the crosswind pushing him off course even though his compass tells him his course is correct. His Doppler radar will, however, alert him to the fact that he is being blown sideways (because of the different Dopplers in his two forward beams). Thus not only does the radar enable the pilot to fly a *level* course, but also, because the radar specifies the amount he should alter the direction of the plane during a crosswind (that is, by "crabbing"), the pilot can accurately navigate regardless of head wind, tail wind, or crosswind.

RADARS FOR CARS

Low-power (low-cost) radars are finding use in automobiles. One system under development utilizes an "adaptive" speed control, controlling both the throttle and the brake. Figure 143 shows a system designed to test the principles involved, installed in the front of a Volkswagen (Fig. 38 shows the radar horns of an actual development model installed on a Lincoln). For these systems, the driver selects a constant speed which the car maintains on an open road. However, when the car's radar detects a second car in front of it, the radar "adapts" the car's actual speed so as to maintain a safe distance between the two cars. This "safe distance" is made larger at higher speeds. The radar measures the range of the car in front and also its relative speed (by means of the Doppler shift).

The use of inexpensive radars is also being explored for "antiskid" applications in cars, involving a concept first utilized in aircraft landing brakes. On icy roadways or airport runways, braking pressure can cause one wheel and not the others to skid, with the result that the stopping force provided by that wheel becomes extremely small. The car or plane then usually swerves dangerously because of the lack of balanced braking.

Fig. 143.   An experimental radar mounted on the front end of an automobile.

The antiskid provision ascertains the rotational speed of each wheel (by means of a tiny radar), and when the speed of one wheel begins to drop below that of the others (that is, just as it is beginning to skid) the speed sensor causes the brake pressure on that wheel to be lowered, thus eliminating its tendency to skid.

Figure 144 shows a tiny radar installed on an automobile for the purpose of accurately measuring the wheel's rotational velocity. Each wheel would be equipped with a similar (very inexpensive) device for detecting wheel-speed changes on initiation of skid. The advent of two inexpensive microwave-generating devices, the Gunn diode and the avalanche diode, played a large part in the development of small, low-cost radars of this kind.

CIRCULAR POLARIZATION

In early radars, as the ability to generate higher and higher micro-wave frequencies appeared, it was noticed that at these very high fre-

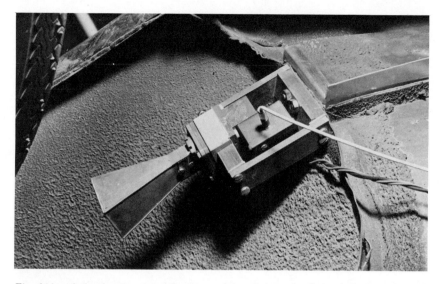

Fig. 144.   A tiny inexpensive radar can measure the rotational speed of each car wheel, thereby helping to provide antiskid performance.   Courtesy Bendix Corporation.

quencies (short wavelengths) heavy rainfall generated such strong echoes that the echoes of the sought-for targets were obliterated by those reflected by the rain. Again, a multidisciplinary mind provided a valuable solution to this problem. Optics experts and many physicists were aware that reflections from objects having circular symmetry and also from "specularly" reflecting surfaces could be reduced if circularly polarized light waves were used to illuminate such objects. But most such experts were not too aware of the latest advances (and problems) of radar. Fortunately, some were cognizant in both fields, and soon experiments were made to test the use of circularly polarized microwaves in a radar to reduce the radar echo from the circularly symmetrical raindrops. This off-the-beaten-track thinking was valuable in enabling such radars to "see through" raindrops.*

This effect can be explained with the help of one of my early metal-plate structures, one comparable to the early metal-plate (waveguide) *lens* structures (such as the one shown in Figure 95). It is the polar-

---

* J. Brown, *Microwave Lenses* (London: Methuen, 1953). W. E. Kock, *Sound Waves and Light Waves* (New York: Doubleday, 1965), p. 71.

ization rotating device shown in Fig. 145* (Appendix 93), where a microwave "half-wave plate" converts vertically polarized waves to horizontally polarized waves. The vertically polarized component of the transmitted waves experiences a phase velocity which is higher than that of the (unaffected) horizontally polarized component, causing the polarization of the emerging waves to be rotated by 90 degrees. If the thickness of this half-wave plate is reduced by one-half, the emerging waves are circularly polarized, and, upon *symmetrical* reflection (such as raindrops), their second passage through the *quarter*-wave plate will again yield the full 90-degree rotation shown in the figure. Because these waves now possess a linear polarization which is oriented 90 degrees to the transmitter–receiver polarization, they cannot enter the receiver. On the other hand, signals reflected from nonsymmetrical objects will experience, on reflection, varying degrees of depolarization;

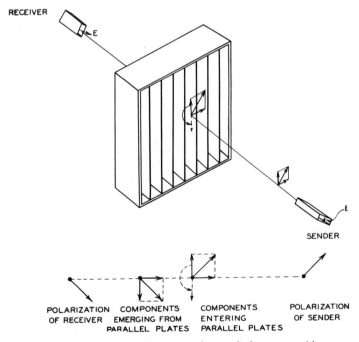

Fig. 145. Rotation of polarization by means of a metal-plate waveguide structure.

---

* W. E. Kock, "Related experiments with sound waves and electromagnetic waves," *Proc. IRE, 47,* pp. 1192–1201 (1959).

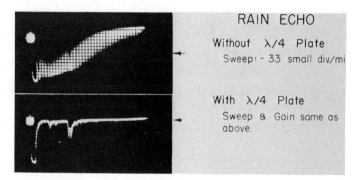

Fig. 146. Suppression of rain echoes in a microwave radar by changing the outgoing signal polarization from linear to circular. The plot is amplitude (vertically downward) versus range (horizontally to the right). Top, linear polarization; bottom, circular polarization.

they therefore will possess a component which *can* enter the receiver (Fig. 146).

Following the early flurry to use shorter and shorter wavelength microwaves in radars (the very short 1.25-centimeter waves were given the designation "K-band"; these were the ones most bothered by rain), a pulling-back took place, and today practically all wavelengths used in military radars are long enough to avoid the rain reflection. But when the *laser* radars (see Fig. 54) began to appear, I thought back on the early (World War II) circularly polarized wave experiments and some diving into the woods took place. In a recent publication, I noted, in connection with Fig. 54, that

> The high detail here recorded, as compared to that obtained by a standard microwave radar, is evident. However, with such optical radars, atmospheric effects (including even light fog) can seriously affect the reflected patterns. If the ratio of the size of fog droplets to rain drops is comparable to the ratio of the size of light wavelengths to radar wavelengths, the use of circular polarization here should yield a result comparable to that shown in [Fig. 146]. This would permit objects otherwise obscured by fog or mist to be made more visible.*

In actual experiments, it was found that coherent light reflected from steam (thereby blocking the view of objects behind the steam) could be eliminated by causing the illuminating light to be circularly polarized.*

---

* W. E. Kock, "Circular polarization in certain laser and holography applications," *Applied Optics* (July 1975).

Who knows, someday we may be using circular polarizers on our own automobile headlights, permitting us to see through fog on the highway, thanks to radar craftiness (Appendix 94).

AN AIRCRAFT-VERSUS-SUBMARINE MACHIAVELLISM

To close this chapter, we describe a very innovative radar technique that proved effective during World War II. In those days, submarines, to function while submerged (that is, at periscope depth), had to operate on their batteries, because their diesel engines required air. This submerged activity thus created a certain amount of discharge to the batteries, and on occasions when they were on the surface they used their diesel engines to recharge their batteries.

During the war, the German submarines performed this recharging operation at night, when the chance of their being seen was, in preradar days at least, quite remote. Then radar arrived, first to the Allies, and significantly later to the Germans (and Japanese). The nighttime-surfaced subs were easy prey for radar (and bomb) equipped aircraft, and many submarines were sunk without any of the personnel being aware that the sub had been detected.

Soon, however, it became known that Allied aircraft were equipped with radar, and soon, therefore, the German submarines were equipped with (passive) listening equipment, designed to reveal that radar signals were being detected at the sub. This gear also indicated the changing strength of the radar signal, thereby informing the submarine commander that the radar was either approaching (the signal then increasing in intensity) or passing by and receding (the signal then decreasing). For the first situation, the sub dove immediately, thereby removing its echoing structure from the water surface and thereby usually making its escape. For the second situation, with the signal strength decreasing, the commander "quite correctly" concluded that the plane had missed the sub and was continuing on, away from it.

The words "quite correctly" are in quotes because as a result of some clever, innovative thinking by a U.S. radar designer (Luis Alvarez, whom we have referred to earlier), the conclusion that the radar was going away (not approaching) soon was an incorrect conclusion.

To understand how this could possibly be true, we first recall that the radar operation involves a two-way path, one the outward path to the sub and the other the echo returning to the plane. For long-range detection of such echos, a very powerful pulse must be sent out, because of (1) the small amount that reaches the sub over the long path, (2) the small amount that is reflected, and (3) the small amount that survives the long trip back. For the short-range situation, it turns out that the radar is way overpowered, because the two paths are then much shorter and the double transit effect is tremendously smaller. The listening sub, however, has only a one-way path to deal with, and its receiver requires only a modest sensitivity because it is listening to the very powerful outgoing pulses of the aircraft radar.

The radar innovator thought about this situation and came up with a brilliant idea to which he gave the name "Vixen": Arrange, once the radar has detected the sub, for the outgoing pulse power to be continually *reduced*. With the two-way path, the rate of reduction with range (that is, as the plane gets closer to the sub) can be so large that, for the sub, its received signal is constantly getting *smaller*. To those in the sub, it seems obvious that the plane is getting farther away. Obvious? Well, obvious until the plane suddenly turns on its searchlights and drops its bomb on the still-surfaced sub.

An example of a valuable contribution by an engineering scientist who left the beaten path of radar engineering and dove into the woods of military strategy.

11

HOLOGRAPHY

In 1958, a revolutionary new optical process called holography was described by the British engineering scientist Dennis Gabor,* (being of Hungarian origin, Gabor maintains that the correct pronunciation of his name places the accent on the first syllable). In this chapter, we discuss some of the ramifications of this Nobel Prize winning concept.

GABOR'S INTERDISCIPLINARY SKILL

Prior to his brilliant exposition of holography, which won for him the 1971 Nobel Prize in Physics,† Gabor had been an outstanding contributor to many varied fields of technology. One very outstanding and very complete paper was of particular interest to me because it (in part) involved some classic deductions about the ability of the ear to determine the pitch of steady, single-frequency tones. In some earlier experiments to test this effect, the *duration* of the tone employed, although quite long and hence quite adequate for the higher-pitched tones, was not sufficiently long for the very low-pitched tones. In a paper which covered many aspects of speech and hearing, Gabor discussed this effect for tones of different pitches, pointing out that short-duration tones, even when produced by a generator of extremely high pitch stability, constitute a *spread* of frequencies, not a single frequency.

* D. Gabor, *Nature, 161,* p. 777 (1948). D. Gabor, *Proc. Roy. Soc.* (*London*), *A 197,* p. 454 (1949).
† W. E. Kock, "Nobel prize for physics: Gabor and holography" (invited report), *Science, 174,* pp. 614–675 (1971).

Now it turned out that Gabor had missed two of my publications, one calling attention to this property of finite-length tones (explaining it on the basis of the classic "principle of uncertainty" as set forth by the famous physicist Werner Heisenberg)* and the other questioning, on that basis, the conclusions earlier published based on experiments in which the tone duration for low frequencies was too short† (Appendix 95). When I alerted Gabor to these earlier papers, he was most apologetic for not including them in his paper's reference list, and from then on he transmitted copies of all of his papers to me. As an indication of the difficulty in comprehending that concept, set forth separately by Gabor and by me, a rebuttal to my papers appeared many years later in the *Journal of the Acoustical Society of America.* Written by S. S. Stevens (a respected acoustics expert), this rebuttal showed that Stevens had not been aware of Gabor's paper corroborating this concept and explaining it in a much more understandable way than my paper had.

HOLOGRAPHY FUNDAMENTALS

Gabor's famous discovery of holography involved an optical process, a *photographic* one, and he named the photographic record which resulted a "hologram," from the Greek words *holo,* meaning "entire," and *gram,* meaning "message." An ordinary photograph collapses its record of the three-dimensional scene photographed into a *two*-dimensional image, whereas the hologram permits a full three-dimensional image of the recorded scene to be "reconstructed." This image is extremely lifelike, and, if the viewer wishes to see around an object in the foreground, he simply moves his head to either side. When illuminated with a special form of light (*coherent* light, as from a laser), the hologram appears as a "window," with a realistic replica of the original scene displayed behind it. Figure 147 shows this ability to look around objects in a hologram. A very spectacular hologram, showing a woman's arm and hand suspended over the sidewalk in front of Cartier's on New York's Fifth Avenue—holding a $100,000 diamond ring and bracelet for all bypassers to see—is discussed at some length in

---

* W. E. Kock, "The principle of uncertainty in sound," *J. Acoust. Soc. Am., 7,* p. 56 (1935).

† W. E. Kock, "A new interpretation of the results of experiments on the differential pitch sensitivity of the ear," *J. Acoust. Soc. Am., 9,* p. 129 (1937).

Fig. 147. Three photographs of a hologram being illuminated with laser light. The hologram recorded a scene comprising three vertical bars; for these photos the camera was moved successively farther to the right, finally causing the rear bars to be hidden by the front bars.

my book *Engineering Applications of Lasers and Holography* (Plenum Press, 1975).

MAKING A HOLOGRAM

In forming a hologram, two sets of single-wavelength light waves are made to interfere. One set is that issuing from the scene to be photographically recorded; almost invariably, it is an extremely complicated one. The other is usually rather simple, often being a set of plane waves. This second set is called the "reference wave," and, in reproducing or reconstructing for the viewer the originally recorded scene, a similar set is used to illuminate the developed photographic plate, the hologram. The two original sets of hologram waves are caused to interfere at the photographic plate, as shown in Fig. 148. Here the "scene" comprises a pyramid and a sphere. The objects are illuminated by the same source of single-wavelength laser light that is forming the plane waves at the top of the figure. Because the wave fronts of the set of waves issuing from the scene are quite irregular, the interference pattern in this case is quite complicated. After exposure, the photographic plate is developed and fixed, and it thereby becomes the hologram. When it is illuminated with the same laser light used earlier as the

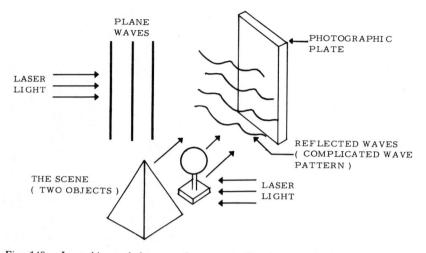

Fig. 148. In making a hologram, the scene is illuminated with laser light, and the reflected light is recorded, along with a reference wave from the same laser, on a photographic plate. The plate is then developed and fixed.

reference wave, as shown in Fig. 149, a viewer imagines he sees the original two objects of Fig. 148 in three dimensions.

To understand how such a light wave interference pattern, once photographically recorded and then developed, can later re-create a life-like image of the original scene, let us first examine the very simple

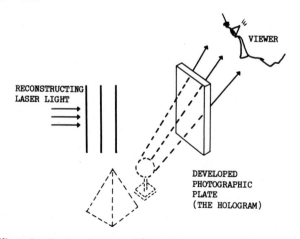

Fig. 149. When the developed plate of Fig. 148 is illuminated with the same laser reference beam, a viewer sees the original scene "reconstructed," standing out in space with extreme realism behind the hologram "window."

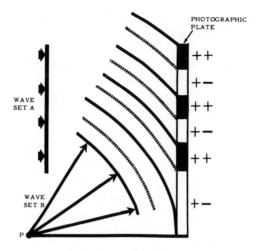

Fig. 150. When two sets of single-wavelength waves, one plane and the other spherical, meet at a plane, a circular interference pattern results, with the separation of the outer circles continually decreasing.

interference pattern formed when a set of plane waves and a set of spherical waves interfere. This pattern is a circular one having a cross-section that is sketched in Fig. 150. Parallel plane waves (set A) are shown arriving from the left, and they interfere at the photographic plate with the spherical waves (set B) issuing from the point source *P*. This interference causes areas of wave subtraction and addition to exist, and, as the distances increase from the central axis, the separation between these areas lessens. The combination of plane and spherical waves generates interference fringes.

A photographic recording of the interference pattern between plane and spherical single-frequency light waves is shown in Fig. 151. Now it turns out that the spacings of the rings of Fig. 150 and 151 are identical to those of a zone plate. The similarity between holograms and zone plates were first noted by the British scientist G. L. Rogers in 1950.* Sections located near the central top and central bottom edges of this pattern resemble somewhat (except for the curvature) the horizontal line pattern of an optical grating.

When an optical grating is illuminated by horizontally traveling, single-frequency plane waves, much of the light wave energy passes straight through the grating. In addition, some of the wave energy is

* G. L. Rogers, "Gabor diffraction microscopy—The hologram as a generalized zone plate," *Nature, 166,* p. 237 (1950).

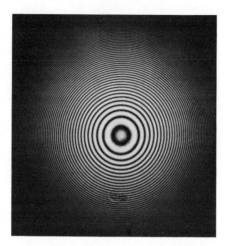

Fig. 151. At the plane of the photographic plate of Fig. 150, the wave interference pattern is a series of bright and dark circles. The figure portrays a photographic recording of such a pattern.

deflected (diffracted) both upward and downward. Accordingly, when the pattern of Fig. 150 or 151 is illuminated with plane waves, three wave sets are similarly generated.

This effect is shown in Fig. 152, which portrays the reconstruction of the hologram formed in Fig. 150. A portion of the reconstructing plane wave set arriving from the left is undeviated, passing straight through the photographic transparency. Because the circular striations near the top of the drawing act like the horizontal lines of a grating, energy is diffracted both upward and downward. However, the pattern of striations is *circular,* so that the waves which are diffracted in the upward direction travel outward as circular wave fronts, seemingly originating at the point P_v. These waves form what is called a "virtual image" of the original point light source P of Fig. 150 (virtual, because in this reconstruction no source really exists there). These waves provide an observer, located where the words "upward waves" appear in Fig. 152, with the illusion that an actual point source of light exists there, fixed in space behind the photographic plate no matter how he moves his head. Furthermore, this imagined source exists at exactly the spot occupied by the original spherical wave light source used in making the photographic record, the hologram.

A third set of waves is also formed by the spherical recorded interference pattern. In Fig. 152, this set is shown moving downward, and these waves are focused waves. They converge at a point that is located at the same distance from the far side of the photographic record as the virtual source is from the near side. The circular striations of the photographic pattern cause a real image of the original light source P to form at P_r (*real, because a card placed there would show the presence of a true concentration of light*).

THE COMPLETE HOLOGRAM PROCESS

The complete, two-step hologram process is shown in Fig. 153. Here, a pinhole in the opaque card at the left serves as the "scene"; it is a point source of spherical waves. These interfere at the photographic plate with the plane waves arriving from the left. In this case, only the

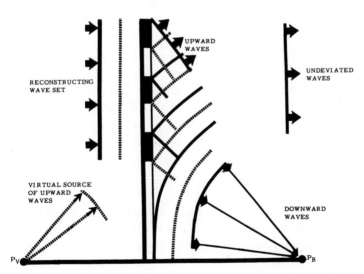

Fig. 152. When the upper portion of the circular line pattern of Fig. 151 (as recorded in the process sketched in Fig. 150) is illuminated with the original, horizontally traveling set of plane waves, three sets of waves result. One set travels straight through horizontally, another acts as though it were diverging from the source point of the original spherical waves, and the third is a set that converges toward a point on the opposite side of the recorded circular pattern.

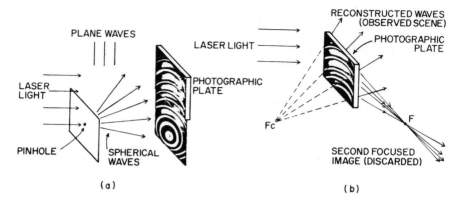

Fig. 153. Plane reference waves interfering with spherical waves issuing from a pinhole form a zone plate interference pattern (a), which, when photographically recorded and reilluminated with laser light, generates waves appearing to emanate from the original pinhole (b).

upper portion of the circular interference pattern is photographically recorded. When the photographic plate is developed and fixed and then placed in the path of plane light waves as shown in the diagram on the right, a virtual image of the original pinhole light source is formed at the conjugate focal point, F_C. A viewer at the upper right thus imagines he sees the original light from the pinhole. The real image (the focused image) appears at the true focal point as shown; this second wave set was the one used in the Cartier jewelry hologram mentioned earlier.* In this figure, the straight-through, undeviated waves are not shown.

In Fig. 154, a similar photographic recording procedure is sketched, except that for this case the original scene is one having, not a single pinhole, but three pinhole sources of light, each in a different vertical location and each having a different axial distance from the plane of the photographic plate. We see that each of the three light sources generates its own circular, many-ring pattern, comparable to the single pattern of Fig. 153. (In the figure, only the first two central circular sections of these patterns are indicated.) The upper portions of the three sets of circular striations (those encompassed by the photographic plate) are recorded as three superimposed sections of zone plates. When this film is developed and fixed and then reilluminated, as was done for

---

* Science, 180, pp. 484–485 (1973).

the single-pinhole recording of Fig. 153, three sets of upward waves and three sets of focused waves are generated. Of particular importance from the standpoint of holography is that a virtual image of each of the three pinholes is generated (by the upward, diverging waves). These virtual images cause a viewer at the top right to imagine that he sees three actual point sources of light all fixed in position and each positioned at a different (three-dimensional) location in space. From a particular viewing angle, source 3 might hide source 2. However, if the viewer moves his head sideways or up or down, he can see around source 3 and verify that source 2 does exist.

Most applications of zone plates exploit only their focusing ability. The fact that a zone plate also causes a set of diverging waves to be generated is not very well known. But in a hologram this negative lens property of a zone plate is very important, for, as has been noted, it is

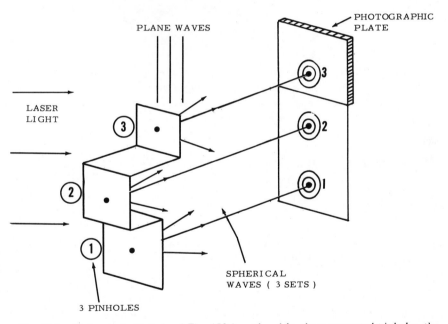

Fig. 154. If the single pinhole of Fig. 153 is replaced by three separated pinholes, the zone plate patterns of all three are photographically recorded; when this photograph is reilluminated, all three pinholes are seen in their correct three-dimensional positions. A more complicated three-dimensional scene can be considered as many point sources of light, each generating, on the hologram plate, its own zone plate; each of these zone plates will then reconstruct its source in its original three-dimensional position.

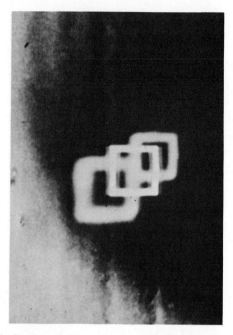

Fig. 155. A 2-inch-diameter hologram zone plate forming three images. The zone plate is the dark area with a circular outline, and the object was a white square. The zero-order (straight-through) image is the bright square; the virtual, diverging image is to its lower left, and the real, converging image is to its upper right.

the diverging waves that give the viewer the striking, three-dimensional view of the original scene. Figure 155 shows the triple image effect produced by the straight-through, the converging, and the diverging waves generated by a zone plate.*

THE HOLOGRAM OF A SCENE

All points of any scene that we perceive are emitting or reflecting light to a certain degree. Similarly, all points of a scene illuminated with laser light are reflecting light. Each point will have a different degree of brightness; yet each reflecting point is a point source of laser light. If a laser reference wave is also present, each such source can form, on a

* W. E. Kock, L. Rosen, and J. Rendiero, "Holograms and zone plates," *Proc. IEEE,* *54,* pp. 1599–1601 (1966).

photographic plate, its own circular interference pattern in conjunction with the reference wave. The superposition of all these circular patterns will form a very complicated interference pattern, but it will be recorded as a hologram on the photographic plate, as was sketched in Fig. 148. When this complicated photographic pattern is developed, fixed, and reilluminated, reconstruction will occur and light will be diffracted by the hologram, causing all the original light sources to appear in their original, relative locations, thereby providing a fully realistic three-dimensional illusion of the original scene.

As noted, the hologram plate itself resembles a window, with the imaged scene appearing behind it in full depth. The viewer has available to him many views of the scene, and to see around an object in the foreground, he simply raises his head or moves it to the left or right. This is in contrast to the older, two-photograph stereo pictures that provide an excellent three-dimensional view of the scene, but only one view. We saw in Fig. 147 three photos of one (laser-illuminated) hologram; they are three of the many views a viewer would see if he moved his head from right to left while observing the hologram. For these three photos, the camera taking them was similarly moved from right to left, fully exposing, in the left-hand photo, the original three bars positioned one behind the other.

PARALLAX IN HOLOGRAMS

In viewing a hologram, the observer is usually encouraged to move his head sideways or up and down so that he may grasp its full realism by observing an effect called *parallax*. In real scenes, more distant objects appear to move with the viewer, whereas closer objects do not. Such effects are very noticeable to a person riding in a train; the nearby telephone poles move past rapidly, but the distant mountains appear to move forward with the traveler. Similarly, the parallax property of holograms constitutes one of their most realistic aspects.

Because hologram viewers invariably do move their heads to experience this parallax effect, hologram designers often include cut-glass objects in the scene to be photographed. In the real situation, glints of light are reflected from the cut glass, and these glints appear and disappear as the viewer moves his head. This effect also occurs for the hologram, and further heightens the realism, as in the hologram of Fig. 156. The silver chalice used in this hologram is one that I acquired

Fig. 156. A photograph of a hologram, showing a cut-glass toothpick holder, a carved silver object, and a "thumbprint" glass. The viewer, as he moves his head, will see changes in the light reflected from these objects' surfaces.

in 1936 in Bangalore, India, during my postdoctoral study with Nobel Laureate Sir C. V. Raman.

SINGLE-WAVELENGTH NATURE OF HOLOGRAMS

One of the basic properties of ordinary holograms (and zone plates) is their single-frequency nature. Because the design of a zone plate is postulated on one particular wavelength, only waves of that wavelength will be properly focused. Inasmuch as holograms are a form of zone plate, they, too, suffer from this problem; only single-frequency light waves can properly reconstruct their recorded images. If light comprising many colors is used in the reconstruction process, the various colors are diffracted in different directions and the picture becomes badly blurred.

NONOPTICAL HOLOGRAMS

Because the basic feature of holography is that of recording a wave interference pattern, holograms can be made using waves of almost any kind. Thus two sets of radio waves, or two sets of sound waves, can be made to interfere, and if records are made of such patterns they will be radio (or microwave) holograms or acoustic holograms.

MICROWAVE HOLOGRAMS

We have seen that, to form a hologram, reference waves are needed. Such a reference for the microwave case can be provided as shown in Fig. 157. Here the microwaves passing through the lens at the left are the waves of interest, and a set of coherent waves from the reference feed horn at the right interfere at the scanning plane. Again a camera can record this interference pattern, and Fig. 158 is the result. The lines seen are interference lines or microwave "fringes," and, as

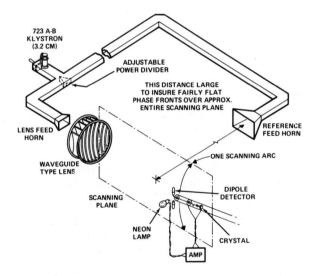

Fig. 157. Two sets of coherent microwaves generated by the klystron at the upper left (one set being an off-angle reference wave) are caused to interfere at the scanning plane. The microwave interference pattern is converted into a light pattern by the scanning mechanism, and it is recorded by a camera set at time exposure.

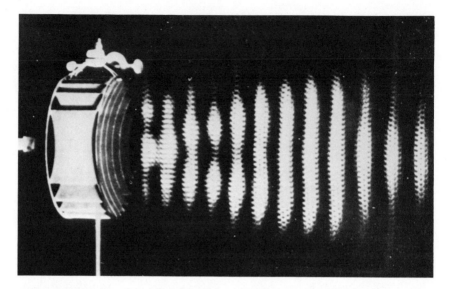

Fig. 158.   Microwaves issuing from the horn at the left are converted into plane waves by a waveguide lens.

has been noted,* this record can thus be classed as a microwave holo-gram. It was made at the Bell Telephone Laboratories in 1951 by me and my colleague Floyd K. Harvey. It was also noted in the excellent text on holography by Kiemle and Ross that this was the first use of an off-axis reference signal to record a ("hologram") interference pattern.*

The telephone receiver pattern of Fig. 159 is also an interference or "fringe" pattern (that is, a hologram), in this case an acoustic one. Because sound waves can be transformed so easily into varying electrical currents (and *vice versa*), a different way of injecting the hologram reference wave was used in making this sound wave record. The process is shown in Fig. 160. A signal coming directly from the electrical oscillator (which is causing the receiver to vibrate and thus radiate sound waves) is combined with the signal picked up by the microphone. At the junction point of these two sources, interferences will occur, and the brightness of the light will again be affected by the constructive and destructive interference effects. The "electronically injected" reference

* H. Kiemle and D. Ross, "Einführung in die Technik der Holographie," *Akad. Verlagsges.*, p. 17 (1969).

Fig. 159. The nondirectional pattern of sound waves issuing from a telephone receiver.

wave is identical in action to that of a set of plane reference waves, and an acoustic hologram results (Fig. 159).

MICROWAVE HOLOGRAMS AND LIQUID CRYSTALS

Microwave interference patterns can also be portrayed using the recently developed "liquid crystal" technique, in which crystals whose colors are determined by the varying temperature effects introduced by

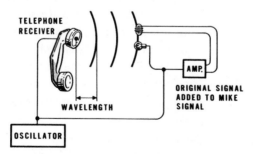

Fig. 160. To make an acoustic hologram, the reference wave can be injected "electronically" by means of a connection directly from the oscillator. Figure 159 was made in the way shown here.

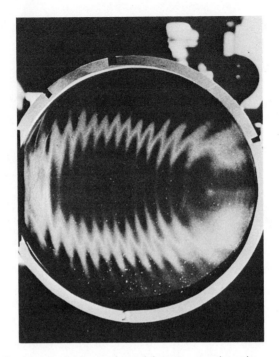

Fig. 161.   A liquid crystal pattern formed between a point microwave source and a plane wave source. Because it is offset, the straight-through (zero-order) component and the two images can be separated.

the variations in the strength of the microwave field are employed. Figure 161 shows such a liquid crystal microwave interference pattern generated by two interfering coherent microwave sources.* One was a point source and one approximated a plane wave source, and the two wave sets were made to interfere at the plane of the liquid crystal device. The resulting pattern is identical to the pattern of an offset microwave zone plate. Since this zone plate is a hologram, the reconstruction of an image (either real or virtual) of the original point microwave source can be accomplished optically by photographically reducing the record and illuminating it with coherent (laser) light.

Had there been a large number of microwave point sources in the original microwave "scene," each would have formed, in conjunction

* C. F. Augustine and W. E. Kock, "Microwave holograms using liquid crystals," *Proc. IEEE, 57,* pp. 354–355 (1969).

with the reference wave, its own two-dimensional zone plate, and the liquid crystal pattern would have been, as in an optical hologram, a superposition of a large number of zone plates. By photographically recording, reducing, and reilluminating this pattern with laser light, a three-dimensional, visual portrayal of the many microwave sources would have been formed.

ULTRASONIC HOLOGRAMS

We have noted that two-dimensional acoustic holograms like the liquid crystal hologram of Fig. 161 can be formed by recording a sound wave interference pattern (Fig. 159). When actual objects comprise the "scene" to be "illuminated" with sound waves, the wave interference pattern is first translated into a light wave pattern that, after a size reduction, is viewed with laser light for optical reconstruction of the scene.

Dr. Rolf Mueller pioneered at the Bendix Research Laboratories in using as an ultrasonic hologram surface a liquid–air interface.* We mentioned this technique earlier in connection with Figs. 59, 60, and 61. A coherent reference wave is directed at the liquid surface, and it becomes the "recording" surface for the hologram interference pattern. Because the surface of a liquid is a pressure-release surface—that is, a surface that "gives" or rises at points where higher-than-average sound pressures exist—the acoustic interference pattern transforms the otherwise plane liquid surface into a surface having extremely minute, stationary ripples on it. When this rippled surface is illuminated with coherent (laser) light, an image of the submerged object is reconstructed. This holographic technique has been especially useful in the medical field, permitting extremely small breast tumors to be detected.

UNDERWATER VIEWING

Acoustic holography when fully developed can become a valuable tool for such underwater viewing activities as naval search and surveillance, detection and identification of schools of fish, and control of harbor traffic. Techniques currently available for these purposes are

* R. K. Mueller and N. K. Sheridan, *Appl. Phys.* (*Letters*), *9*, p. 328 (1966).

limited to sonar and optical viewing. Optical techniques are limited, sometimes severely, by the turbidity of the water.

EARTH EXPLORATION

The application of holographic techniques to large-scale underwater and underground viewing is still in initial development stages but promises to aid in the accumulation of geological information for scientific purposes and to have important implications for the mining and oil industries. Object sizes in work of this type are several orders of magnitude greater than those in the ultrasonic applications just discussed, and object distances are measured in hundreds of meters. For good penetration of the geological structures, the frequency range employed must be between 10 and 100 hertz. Frequencies so low (and wavelengths so great) place unusual constraints on detector arrays, but scanning techniques could be applicable in this work, since the objects under study are immobile.

The basic elements of a system for locating offshore oil deposits are depicted in Fig. 162. A cable which in practice would be 100 wavelengths or more in length is towed behind a ship equipped with a high-power transmitter capable of emitting low-frequency coherent acoustic energy into the ocean depths. Signals reflected or scattered from the ocean bottom or from the geological layers below are picked up by the

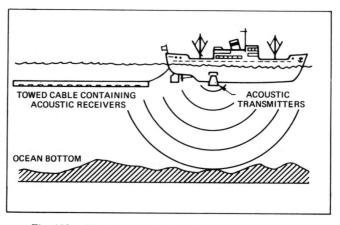

Fig. 162. Underwater viewing using acoustic holography.

cable array. Holographic processing of seismic data obtained from hydrophone arrays should permit the retrieval of useful information from such acoustic signals.

One fairly recent seismic holography experiment was carried out on the Gulf of Mexico over a known and well-defined salt dome. This work was a joint venture of the Bendix Research Laboratories, Bendix United Geophysical Corporation, and Amoco Production Company, previously known as Pan American Petroleum Corporation. A prior experiment in Lake Erie had been successful. The test site for the more recent experiment was in the Eugene Island block 184 field, off the coast of Louisiana, some 70 miles south-southwest of Morgan City (the location of the salt dome). A fairly complete account of this test has been published.* The diameter of this salt dome is approximately 2 miles, and the top of the dome is at a depth of 1156 feet. The experimental data collected permitted the construction of holograms. Linear holographic information was processed, with the linear holographic sampling aperture being approximately 12,000 feet long, extending from the point near the top of the salt dome out over the edge of the dome. The aperture was irradiated using a separate hydraulically powered seismic vibrator for each frequency (driven from its electronic reference source) mounted on the ship. Seismic energy reflected and scattered from the layered structure and scattering centers in the earth was detected with piezoelectric sensors mounted in a cable towed behind the ship, the cable being towed at a depth of about 30 to 40 feet. Figure 162 is a rough sketch of this type of equipment. Holographic data were obtained using three discrete frequencies of 25, 42.2, and 70.4 hertz. Sampling intervals along the aperture corresponded to 1/8 wavelength of the signal in water, which for 42.8 hertz is about 15 feet.

The signal-processing procedure consisted of mixing the received signal with a reference signal, the reference signal being the same signal used to drive the vibrators, and integrating the result for a period of time corresponding to a ship's movement of 1/8 wavelength. Two signal values were recorded for every sampling plane, the second signal being the received signal mixed with the reference wave delayed by 1/4 wavelength. This procedure provides a value which is in phase quadrature with the first. By using both of these values in the holographic reconstruction, an unambiguous phase with respect to the reference was

* J. B. Farr, "Earth holography to delineate buried structures," *Acoustical Holography*, Vol. 6, (New York: Plenum Press, 1975).

obtained. The outline of the dome in its expected position was found in the hologram reconstruction.

PHASE QUADRATURE

The clever phase quadrature technique just described is largely due to J. S. Keating of the Bendix Research Laboratories. Its ability to achieve a very sizable improvement in hologram efficiency reminded me of a microwave antenna technique that I employed during World War II (following the development of improved transmission in optical lenses through the use of "coatings"). Appendix 96 shows the use of quarter-wavelength (phase quadrature) steps utilized on a microwave *lens* (comparable to the coating on an optical lens). In such lenses, wave energy *reflected* from the quarter-wave low spots is 1/2 wavelength out of phase with that reflected from the high spots. Figure 163 shows this canceling effect for two opposite-phase waves. This *reminding* suggested me that Keating's technique for holograms should also be able to improve the efficiency of the zone plate, one of the simplest holograms.

From time to time, zone plates have been experimented with, particularly in the microwave field. E. Bruce incorporated a zone plate in the antenna of a World War II radar,* and Fig. 164 shows a microwave zone plate antenna investigated by I. Maddaus, Jr., at the U.S. Naval Research Laboratory.† However, an inspection of Fig. 152 shows that half of the incident energy has to be reflected (not shown in the figure) because half of the area of the zone plate is opaque (reflective).

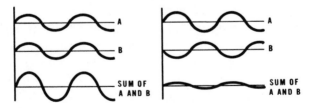

Fig. 163. Two waves of the same wavelength add (at the left) if their crests and troughs coincide and subtract (at the right) if the crest of one coincides with the trough of the other.

---

* E. Bruce, "Directive radio system," U.S. Patent 2,412,202 (December 10, 1946).

† I. Maddaus, Jr., "Fresnel zone plate antenna," *Naval Research Laboratory Report R-3293* (June 1948).

Fig. 164. A microwave zone plate.

Further, energy passes straight through (shown as the "undeviated waves") and the remaining energy divides equally between the desired "downward waves" and the undesired "upward waves." One sees that only a small amount of energy is focused at the real focal point, in contrast to a lens where most of the energy is concentrated at the one (real) focal point.

So here is where Keating's phase quadrature technique enters the picture for microwave zone plates. My paper* suggests a *double* zone plate, with the two units separated by 1/4 wavelength, and both designed to focus waves at the same focal point. These two different zone plates could cause the reflected waves to almost completely cancel. Furthermore, the two zone plates would not have the proper design for the same "upward waves" (or for the "undeviated waves") of Fig. 152. Only the *focused* set of the six wave sets would be affected properly by the two zone plates, resulting in much more energy being focused with the double zone plate than with the standard (single) one.

---

* W. E. Kock, "Parallel optical computing," invited paper presented at the 1976 International Optical Computing Conference, held in Capri, Italy (August 29–September 3, 1976).

LASER HOLOGRAPHY

In 1948, when Gabor described this new principle, the first laser had not yet been built, so that the available sources of the needed coherent light were rather low-intensity ones. Although Gabor's experiments definitely proved the validity of the hologram concept, interest in his discovery was, for many years, rather limited. We mentioned earlier* that scientist G. L. Rogers recognized the importance of Gabor's concept and wrote several important publications during this period. His famous "zone plate paper" of 1950 made the explanation of hologram theory very simple, and his less well-known 1956 paper,† suggesting the *radio* form of holography, was the first to describe the addition of the coherent transmitted radio signal as the hologram reference wave, thereby causing the radar to acquire a higher "synthetic" gain (it later became a "synthetic aperture").

It was not until 1963 (following the demonstration in 1960 of the first laser by the U.S. engineer T. H. Maiman,‡ then at the Research Laboratories of the Hughes Aircraft Company) that the eminent U.S. engineering scientist Emmett N. Leith (Fig. 165) introduced the laser to holography. The subsequent advances made by him, by another U.S. scientist George W. Stroke, and by their many co-workers (all at the University of Michigan at that time) led to a tremendous explosion in holography development.

THREE-DIMENSIONAL HOLOGRAPHY

In viewing a hologram, the three-dimensional illusion is far more realistic than Fig. 147 can convey. The viewer quickly realizes that much more information about the scene is furnished by a hologram than by other three-dimensional photo processes, such as stereophotography (involving a pair of stereophotos). In the hologram reconstruction, the viewer can inspect the three-dimensional scene not just from one direc-

* Rogers, *op. cit.*

† G. L. Rogers, "A new method of analyzing ionospheric records," *Nature, 177*, pp. 613–614 (1956).

‡ T. H. Maiman, "Stimulated optical radiation in ruby," *Nature, 187*, pp. 493–494 (1960).

Fig. 165.   Emmett Leith, the first to use lasers in holography, illuminates one of his holograms at NASA's Electronic Research Center. Courtesy NASA.

tion, as in stereophotography, but from many directions (only three of which are shown in Fig. 147).

As early as 1949, Gabor (Appendix 97) wrote that "the photography contains the total information required for constructing the object, which can be two-dimensional or three-dimensional."* For his two early papers, Gabor used two-dimensional transparencies for his objects, and Leith, using a laser for the first time,† likewise employed two-dimensional transparencies, later, however, contributing extensively to three-dimensional holography (for example, Fig. 165), taken at the NASA Electronics Research Center when I served for 2 years as its first Director).

George Stroke (Appendix 98), a recognized authority on the ruling of precision optical gratings, commenced quite early looking upon holograms as diffraction devices, and in a May 1964 set of lecture notes

* D. Gabor, *Proc. Roy. Soc. (London)*, *A197*, p. 454 (1949).
† E. N. Leith and J. Upatnieks, *J. Opt. Soc. Am.*, *53*, p. 1377 (1963).

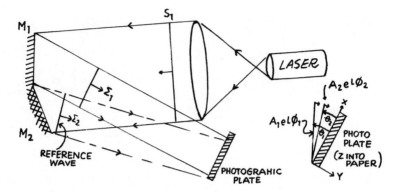

Fig. 166. A sketch from George Stroke's 1964 lecture notes at the University of Michigan.

distributed by his students at the University of Michigan* he described how a three-dimensional object would generate a three-dimensional hologram image (Fig. 166).

INFORMATION CONTENT

The property of holograms which provides the viewer with so much information about the scene recorded is also the property which has limited the use of hologram principles in many interesting applications. Such applications include three-dimensional movies and three-dimensional television, fields which would obviously benefit if the realism of holograms could be imparted to them. The large information content of a hologram is inherent in the extremely fine detail which must be recorded on the hologram film (about 1500 lines per millimeter). Present television systems employ a far, far coarser line structure, so that the outlook for using holograms in television is bleak.

Several methods have been suggested for reducing the information content of a hologram without completely sacrificing its interesting properties, but this task is a very difficult one. Television pictures in the United States have approximately 500 lines vertically and 500 dots horizontally. One television picture (corresponding to one frame of a movie film) thus has an information content corresponding to 500 × 500 or

---

* Now largely included in the well-known book by G. W. Stroke, *An Introduction to Coherent Optics and Holography*, 2nd ed. (New York: Academic Press, 1969).

250,000 dots. A 200- by 200-millimeter hologram (approximately 8 by 8 inches) having 1500 fringes (lines) per millimeter would have the equivalent of 200 × 1500 or 300,000 lines vertically and 300,000 dots horizontally, with a total information content corresponding to 90,000,000,000 dots (90 billion dots). The ratio between the information in an 8-inch square hologram and that of a U.S. television picture is thus 360,000. To reduce the information content of a hologram by a factor that large would be quite an accomplishment.

Several procedures have been proposed to permit at least some reduction in the hologram information content, including one that I suggested in a February 1966 publication which involved retaining many tiny areas (perhaps 300 to 500 in each vertical line and a similar number in each horizontal line) and discarding the remaining much larger areas.* Each retained area, although extremely small, would still contain several light fringes, and the assemblage would therefore still possess the zone plate character of the original hologram. Appendix 99 is the first page of my 1970 U.S. patent on this concept.

THE CONCEPT OF PHASE IN HOLOGRAPHY

In phased-array systems, the concept of phase is quite useful, both in sonar and in radar. When an array is "phased" in one particular way its beam points in one direction, and when phased in another way the beam points in another direction. One procedure often followed in the case of passive receiving arrays is that of incorporating large numbers of elemental delays or phase shifts. The individual element receivers are connected in a multitude of different ways, so that many receiving beams, all pointing in different directions, are effective at all times. The array then "looks" in many directions at once.

Because a photographic plate of film is sensitive only to the intensity of the light falling on it, it is often said that ordinary photographs record only the intensity of amplitude, not the phase of the light field, and *strictly* this is quite correct. Because the radar and sonar engineer utilizes phase and phase shifts as he "phase steers" his arrays, he normally looks upon a receiving array that can thereby "look" in many directions at once as taking phase information fully into account.

* W. E. Kock, "Hologram television," *Proc. IEEE, 54,* p. 331 (1966).

The phasing of his array elements must be done completely and precisely in order to provide all beams with maximum effectiveness.

Now a passive sonar or radar "looking" in many directions is practically identical to a camera that looks in many directions at once, recording on its film the light strength arriving from all these directions (just as the passive sonar measures the sound intensity arriving from many directions at once). But as we noted above the camera film can record only the intensity of the light pattern generated at the plane of its emulsion; it does not record the *relative* phases of various separated areas of the pattern. Accordingly, a further refinement of the definition of phase is needed.

Let us think once again about the light pattern at the plane of the emulsion of the camera film. It unquestionably does have variations in intensity, which provide through the usual photographic process a picture of the scene. But there is also phase information in that light pattern. For example, the waves responsible for two different but equally bright points in the final picture might have been different in their (relative) phase by a full 180 degrees, yet both would appear as indistinguishable bright spots in the final picture. A similar out-of-phase situation could exist in the signals received on two sonar beams without affecting in the least the final sonar record.

But the fact that phase variations which are not detectable by radars, sonars, and photographic plates can and do exist in the pattern does not yet tell us that this information is important. It took holography to show how useful a record would be that *did* record both the amplitude and the (relative) phase of the wave pattern. First, a way had to be found to record these two features of the pattern, and Gabor achieved this by his use of a reference wave. The plane reference waves can clearly indicate the situation in the example above, since the one bright area having its phase 180 degrees opposed to the other would interfere differently with the reference waves. Obviously, the three-dimensional reconstructions of holograms, not possible with one ordinary photo, proved without a doubt that the phase information is useful and that photography *cannot* lay claim (as holography can) to providing the "complete message."*

---

* W. E. Kock, *Lasers and Holography* (New York: Doubleday, 1969).

SYNTHETIC APERTURE RADAR

One of the most extensive uses of nonoptical holograms has been in the microwave radar field in the form of synthetic aperture radar. In 1971, shortly before the announcement of his winning of the 1971 Nobel Prize in Physics for holography, Dennis Gabor commented:

> Unknown to me, a most interesting branch of holography was developing from 1965 onwards at the Willow Run Laboratory attached to the University of Michigan. It was holography with electromagnetic waves, and reconstruction by light, which was called "Side Looking Radar" or "Synthetic Aerials." It was classified work; the first publication by Cutrona, Leith, Palermo and Porcello occurred in 1960. Reconstructions of the object plane by illumination with a monochromatic mercury lamp were of impressive perfection.*

Because Leith was one of the principal contributors to synthetic aperture development, and because he also was with Upatnieks the first to use the laser in holography, Gabor has commented: "Emmett Leith arrived at holography by a path just as adventurous as mine was. I came to it through the electron microscope, he through side-looking, coherent radar."

Thus it is a matter of record that whereas Gabor's brilliant conception of optical holography in 1947 lay almost dormant until Leith initiated in 1963 what has now become a massive worldwide development effort on it, the electromagnetic form of holography received extensive attention all through the 1950s. As noted by Gabor (above) the airborne form of synthetic aperture radar was pioneered by Cutrona and his group at the University of Michigan† and sizable support was given to it in classified projects in numerous laboratories across the United States. Indeed, it is likely that this early classified program was in large part responsible for the extreme rapidity of growth in the development of true optical holography from 1963 on, particularly at the University of Michigan.

---

* D. Gabor, in *Optical and Acoustical Holography*, ed. by E. Camatini (New York: Plenum Press, 1972).
† L. J. Cutrona, E. N. Leith, C. J. Palermo, and L. J. Porcello, "Optical data processing and filtering," *IRE Trans. Inform. Theory, IT6*, p. 386 (1960).

TWO RECENT HOLOGRAPHY DEVELOPMENTS

We discuss here briefly one rather interesting recent application of holography and a second quite significant recent development in the medical field.

In 1975, researchers at Siemens (Germany) developed a laser-based soldering technique that allows a number of component connections to be made simultaneously on a printed-circuit board in a one-step process.* First, a large-diameter laser beam is applied to a hologram, which splits it into several small-diameter partial beams. These fragments are then focused onto the spots to be soldered. The hologram determines into how many beams the original one is to be split and fixes the spots onto which they are to be focused. The technique can be envisioned by recalling Fig. 154; a number of "pinholes" would be used in constructing the hologram, and the *real* images of these holes (as in Fig. 153) would cause a concentration of laser energy at a number of points.

More recently, hologram pioneer George Stroke described a new hologram technique for examining interior structures of the human body.† The method combines both sonic and optical techniques, beginning with a series of conventional, two-dimensional sonograms made of adjacent planes each of which can resolve a plane as little as 0.5 millimeter thick. Then a conventional, optical hologram is made of the first sonogram on a piece of film. Next the first sonogram is replaced in the holograph's laser beam with the second, displaced by the same distance as the separation of the planes of the original sonograms. It is photographed on the same piece of film, which has been displaced by the same distance. This procedure is repeated for all the sonograms, which then appear simultaneously, with their original separation, when the film is developed and viewed by laser light.

Innovations in optics thus continue, with inventive engineers providing new ideas for *using* the discoveries made by others.

GABOR'S CLAIRVOYANCE

We close this chapter with a few observations concerning the depth of thought involved in the invention of holography.

---

* *Electronics,* p. 56 (June 12, 1975).
† George Stroke and Gilbert Baum, *Science* (September 19, 1975).

One of the most intriguing aspects of holography is the way that all of the parts of the quite complicated puzzle came together so elegantly for Gabor. It suggests an inventiveness and a clairvoyance not often encountered.

Gabor's first step was to recognize that all properties of a wave pattern formed at a plane (for example, the plane of a window) by single-wavelength light issuing from a scene are completely specified once the size and relative positions of wave crests and troughs (the amplitudes and relative phases) are specified. (Even Gabor's basic hypothesis was questioned by some authorities on optics.) This recognition was, however, just the first step; to exploit it, he had to find not only a way of recording both amplitude and crest position (phase) of the pattern but also, far more important, a way of retrieving from such a record the original wave pattern. Gabor recorded his wave patterns photographically, and his hologram records did indeed, therefore, contain the required amplitude and phase information about the wave pattern. For the information to be *retrieved,* however, so as to generate a replica of the wave pattern, a third step was required—one far from obvious. In recent years, the problem of usefully retrieving information which has been recorded (as, for example, in thousands of books in a library) has been attacked with great vigor. If someone had asked an information-retrieval expert to retrieve the information contained in the 36 billion dots of one hologram record he would have looked upon that task as a very complicated one. Had he been told that the simple procedure of shining coherent light on the record would do the job, he would surely have been most skeptical, *until,* that is, he witnessed the unbelievable three-dimensional image.

Gabor set down two equations (which Stroke has called "the two basic Gabor equations") which showed that phase and amplitude *would indeed* be recorded. But they also showed that there would not be a one-to-one correspondence between the hologram record and the desired wave set (we noted in connection with Fig. 153 that a real image *and* a virtual image were generated). Also, if the blocking zones of Fig. 152 had been reflecting rather than absorbing, plane waves arriving from the left would have generated, in addition to the three wave sets (converging, diverging, and straight-through) on the right, a similar group of three additional reflected wave sets on the left. Accordingly, if a wave set similar to any one of these four (reconstructed) converging and diverging wave sets had been used in the original recording process (in conjunction with a plane reference wave set), the identical (planar) hologram zone plate would have resulted. The situation is comparable to that of recording a word which has numerous meanings, and then trying to avoid ambiguity in the replay and ascertaining which of the meanings was really intended in the recording.

The ambiguity could easily have daunted another investigator, but it did not daunt Gabor. When he set down his second equation, expressing mathematically the interaction between the recorded pattern (the hologram) and a third wave set (actually, one identical to the reference wave), he realized that wave sets other than the desired set would also be reconstructed. Nevertheless, the fact that the desired set would indeed be reconstructed was of far greater significance.

Gabor subsequently proved experimentally the validity of his conclusions, but in order to achieve maximum image brightness with his weak source of coherent light he viewed his images by looking directly toward the source; the resulting superposition of images caused some blurring. Sixteen years later, Leith was able, with the far more powerful light from a laser, to exploit the offset procedure of coherent radar, thereby angularly separating the several components. The puzzle was assembled, the mission was accomplished, and our hologram story ends.

12

PICTUREPHONE

In this chapter, we discuss the beginnings and the later developments of the Bell System equipment which permits a telephone user to *see* the person he is talking to, equipment referred to by the copyrighted name "Picturephone."

As early as June 3, 1906, the *American Weekly* carried a story on a "seeing telephone" called the "televue" (see Appendix 100), which, according to the article,

> will revolutionize the conditions of modern life, perhaps even more completely than the telegraph or telephone. . . . With the simple telephone it is difficult to buy goods with confidence because naturally one wishes to see them before buying.
>
> But this difficulty will be removed by the televue. It will simplify the task of the housewife enormously. She will be able to buy dress goods and provisions and do all her shopping by televue. The salesgirl will hold up the article desired before the televue transmitter and say: "How will this hat suit Madame?"

And in an article entitled "Picturephone to Change Its Image," in the September 13, 1973, issue of *Electronics,* a photo appeared showing a woman shopping for dresses while having her hair done. Since some form of this service may someday become as commonplace as the telephone is in today's world, let us review how some of the concepts arose which led to the Picturephone.

VISIBLE SPEECH

We discussed in Chapter 3 the development of the "visible speech" technique by Bell Laboratories engineer Ralph Potter (Potter was one

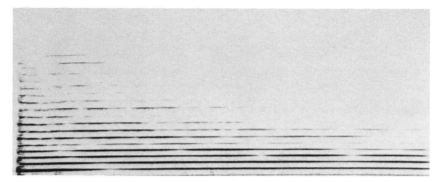

Fig. 167.   Visible sound analysis of a piano tone.

of the department heads to whom I reported and to whom I dedicated my book *Seeing Sound,* Wiley, 1971). In the visible presentation (such as we saw in Fig. 18, 19, and 20), time is plotted horizontally, and frequency vertically, with the loudness of the sound being indicated by the shade of darkness of the record. Thus in Fig. 167, which portrays the changing pattern of the many harmonics of a piano tone when one piano key is struck and held down, it is seen that the higher harmonics die out first. In Fig. 168, the loudness of a sound which has no harmonics is not changed, only its pitch is varied. We saw earlier how this analysis procedure was able to portray the important *formant* characteristic of the human voice and of some musical instruments.

Now the original visible speech equipment required that the sound to be analyzed first be recorded, with a much longer time then being

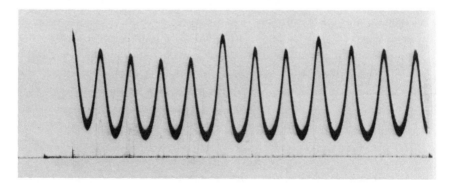

Fig. 168.   A tone varying in pitch.

required for the generation of the visual analysis. During the early 1950s, a new, "real time" technique of sound analysis was developed by Potter and his associates. Because it was this new technique which led to the early Picturephone experiments, let us review this development.

REAL TIME SOUND ANALYSIS

In the newer analysis technique, an observer can *see* the spectrum analysis of the sounds almost as the sounds are being generated. For this device, Potter and three of his colleagues, Larned Meachum, Harold Barney, and Ira Cole, chose to use a tape recording and playback procedure to achieve the speedup in time. The sound to be analyzed was recorded on relatively slowly moving magnetic tape (not unlike the procedure used in today's standard tape recorders), but the playback was made quite fast by causing the tape to move at extremely high speed past the magnetic playback head. As is the case when the tape on a standard tape recorder is moved ahead rapidly (for example, in the "fast forward" condition), the recorded sounds are thereby reproduced at a much higher frequency. Music or speech then sounds very odd and, of course, much higher pitched than the originally recorded version does.

In the real time device, the high relative motion of the tape past the reproducing head was accomplished, as shown in Fig. 169, by moving

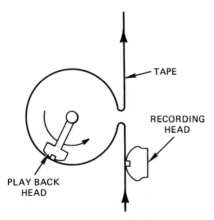

Fig. 169. A tape record–reproduce or record–playback system, where the recording takes place with slowly moving tape and playback occurs with a rapidly moving reproduce head.

the *head* rather than the tape. After the slowly moving tape passed the original recording head, it was made to pass into a structure which formed it into a circular loop. The reproducing or playback head was placed on the end of an arm attached to a rapidly rotating shaft located at the center of the circle of the tape loop. The reproducing head thereby moved in a circle, constantly in contact with the tape loop, except for the small gap in the circle where the tape entered and left the circular structure.

Effectively, then, the section of tape in the tape loop was played back, over and over, with the effective number of playbacks corresponding to the ratio of the rapid playback speed (reproducer head velocity) to the slower, recording tape speed. Very high ratios were possible (far exceeding 100), so that the reproducer, and its following spectrum analyzer, had many, many "looks" at the same piece of tape before a completely new section moved slowly into place.

THE REAL TIME ANALYZER

The spectrum analyzer, as in the original Potter spectrography, comprised a sharp (narrow-band) filter. For each "look," or each sweep of the playback head around the tape loop, the position of the filter in the frequency scale was shifted (usually by an amount equal to the bandwidth of the filter). If the tape in the loop had a sound of that frequency recorded in it, there would be an output from the filter, the magnitude corresponding to the strength of that frequency component. Figure 170 depicts a spectrum analysis (frequency analysis) of a periodic

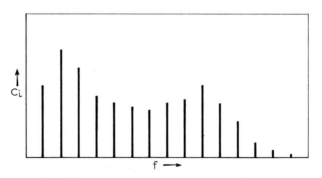

Fig. 170.   A frequency analysis of a periodic sound.

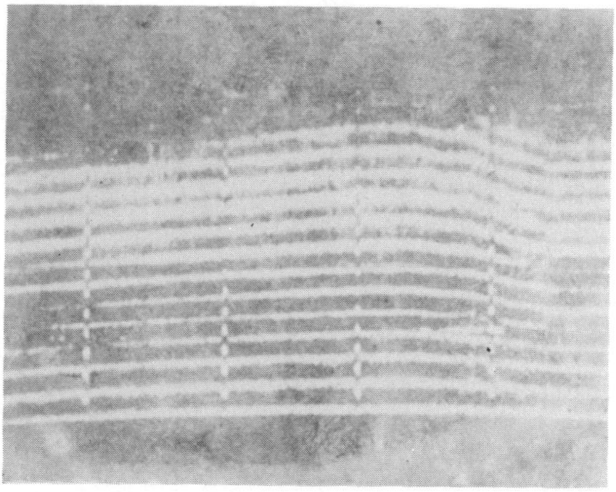

Fig. 171. Analysis of the changing speed of a diesel engine.

sound recorded on the tape. To picture how the analysis was made, let us imagine the filter of the analyzer just discussed to be moving from left to right in this figure. When its position in the frequency scale is such that it is between two of the vertical lines shown, there is no sound (no frequency component) and hence there is no output from the filter. When it is positioned so as to be lined up with, say, the line farthest to the left, a strong signal is generated in its output, and that line is therefore tall (high on the amplitude scale shown at the left). Similarly, if the filter is lined up with the lines at the far right, the output is much less, and those lines are therefore shorter. The plot of Fig. 170 is referred to as variously as a "spectrum analysis," a "frequency analysis," or, after the famous French mathematician, a "Fourier analysis" of the particular periodic sound recorded. Figure 171 shows a real time analysis of a diesel engine. In this record (similar to the record of Fig. 167), the engine speed is first increased and then decreased. This effect is evident in the tilted lines corresponding to the positions of the harmonic components of the diesel engine.

FROM ANALYZER TO PICTUREPHONE

We now come to the spinoff which resulted from this development. I and my associates considered at some length *other* uses of the technique of slow recording and fast playback. This "diving into the woods" resulted in a concept in which the real time technique was used, not for sound, but for *television*—specifically, as a means of providing an inexpensive form of telephone television or "Picturephone," as it is now called. In this early form, the picture was envisioned, not as a continuous, high-resolution picture, as we see on standard television, but rather as one of much lower resolution and therefore one involving a much lower transmission cost. Television, by then, had become fairly commonplace, and full "closed-circuit" television was rapidly recognized as one means of providing a television picture link between two phone users. Only the high cost posed a problem.

TRANSMISSION COST

It is a fair rule of thumb that the transmission cost for telephone or television is proportional both to the distance involved (for example, for long-distance phone calls) and, what is far more important in considering telephone television, to the *bandwidth* (frequency spread) of the two signals involved. We are all aware of the fact that two independent phone lines cost roughly twice as much as one line, but phone circuits require only about 4000 cycles bandwidth for reasonably good intelligibility, whereas a standard television signal requires over 4 million cycles of bandwidth, equivalent therefore, very roughly, to about 1000 times difference in cost. With such a high cost, one might wonder how network television can afford to exist; here, however, the much larger television audience makes it attractive for advertisers to pay for this high cost of transmission. For the telephone user, however, the high cost differential would obviously make him think twice before deciding to add Picturephone to his telephone.

Hence, to minimize this huge cost differential between voice transmission and full television picture transmission, the possibility of using a much-lower resolution picture, by way of the real time sound analysis technique, was explored. It was ascertained that if a picture comparable to a small newspaper (halftone) photo were acceptable, and if it were transmitted not continuously, as in normal television, but only once every 2 seconds, then the entire Picturephone signal could be contained in *one single telephone voice channel,* that is, in a channel encompassing 3500 hertz (3500 cycles per second). The cost to the user would thus be only about double his normal phone cost. To explore whether such a sequence of still pictures of the talker would be of interest, the first step in this program was to make a movie in which this effect was simulated.

THE EXPERIMENTAL MOVIE

In designing the movie to test reactions to a two-way phone conversation in which the picture portion is a series of "stills" and in which a new picture is presented only after 2 seconds has elapsed, numerous concepts were introduced. It was decided that the viewed person should be a secretary talking with her boss and that she would show him her new hair style by turning her head (which could, and *did*, cause a blurred series of five pictures) and also that she could hold up items to the camera and ask questions about various items. The phone conversation went as follows:

Secretary: "Hello, Mr. Kock."

Mr. K.: "Oh, hello, I see you've arrived in San Francisco already."

Secretary: "Yes, my brother met me at the airport, I brought the microphone along. [Holding it before the camera lens] Is this the right one?"

Mr. K.: "Yes, that's just what we wanted. That's swell. Say, haven't you fixed your hair differently, or something?"

Secretary: "Yes, I have. It's different [turning her head], especially in the back. Do you like it?"

Mr. K.: "Very much. But to get back to business, did you find the location of that office here that Roberta wanted?"

Secretary: "Yes, Mr. Smith drew a map of it for me. [Holding it up] Here it is. Maybe you'd like to make a copy of it. Fourth entrance on Oak Street."

Mr. K.: "Oh yes, I see. Well, I should be able to find that all right. Now, did you get to see Mr. Green yet about that letter he sent us?"

Secretary: "I think I brought the wrong letter. You remember there were two Carl Greens. Here is the signature of the one I brought. Is this the wrong one?"

Mr. K.: "No, that's the right one all right."

Secretary: "I certainly am glad to hear that. This Picturephone is quite a gadget. You people at Bell Labs should do recruiting—should do your interviewing over this thing."

Mr. K.: "Well, could be. Of course, then the recruiters might miss those nice trips to California. Well, thanks for calling, and I'll see you next week, then."

Secretary: "Goodbye." [End of film clip.]

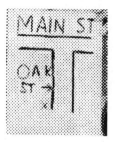

Fig. 172.   Low-detail map for transmission via Picturephone.

To illustrate various portions of the film, Fig. 172 shows the simple street map (Oak Street), to indicate that even rough-detail drawings, the only ones that can be transmitted over the low-detail circuit, can be useful. Fig. 173 shows the "Carl Green signature" (also low detail), and Fig. 174 shows the blurred picture generated when the secretary turned her head to show off her new hair style.

This movie was shown to many persons at the Bell Laboratories and the large majority expressed the opinion that the 2-second, still-picture feature was not objectionable, for a rather interesting and very important reason. Because the telephone *speech* was fully continuous, its continuation covered the gap, and the viewer felt that he was really seeing—live—the person shown in the Picturephone movie (the secretary). The blurred picture was also found not objectionable because it was quickly replaced with an unblurred one. The continuing telephone speech, such as during short periods of laughter, gave a strong illusion that the "program" being observed was a close equivalent to "live television." On the basis of the reactions of these viewers of the movie, the decision was made to implement, in an experimental way, a Picturephone based on the movie.

Fig. 173.   A low-detail signature for transmission via Picturephone.

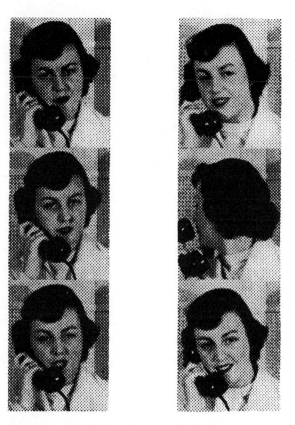

Fig. 174. A 2-second picture is blurred through motion.

THE FIRST PICTUREPHONE

Using the two criteria of low picture repetition rate and low picture detail, numerous ways in which the real time sound analysis procedure could be used were examined. The basic process itself looked attractive in almost its original form, and a system using the tape recording and playback procedure of Fig. 169 resulted. Pictures of the quality of the one shown in Fig. 175 were transmitted, and this detail and the 2-second repetition rate seemed adequate (considering the low cost).

By reversing the record and playback heads shown in Fig. 169, the picture signal was recorded very quickly from a camera tube on magnetic tape. The picture thus corresponded to a short-exposure photo.

Fig. 175.   Detail possible in the first Picturephone.

Figure 176, taken from a memorandum I wrote in 1954, shows the rapid motion of the record head on the tape loop and the slow tape movement past the pickup head. Thus the recorded signal tape moved slowly across the tape playback head at a speed such that the (moderately) high-detail picture signal (the signal corresponding to one picture) would take 2 seconds for the playback (that is, for the transmission) process. At this slow speed, the normally *high-frequency* picture information recorded during one circular sweep of the rotating head was reduced to a much lower-frequency band, one capable of being satisfactorily handled by the standard 3500 cycles per second telephone circuit.

At the receiving end, the procedure shown in Fig. 169 was again used. The 2-seconds-long, reduced-speed, low-frequency signal was recorded on the slowly moving magnetic tape, as was done in the real time sound analyzer. The tape was then stopped, permitting the rapidly rotating playback head to "play back" the now high-frequency picture signal into the viewing tube. While that 2-second signal was being recorded (at the lower left of Fig. 176), a previously received 2-second signal was being played back (at the lower right of Fig. 176) on another similar device, again by a high-speed rotating reproducing head, over and over, quite rapidly. The result for the viewer was a steady (non-flickering) picture on the receiving end viewing scope (the TV tube).

THE PERMANENT RECORD FORM

A second procedure for viewing the transmitted picture was the recording of the slow-speed signal on paper which responds to electrical signals by becoming black when the signal is strong, becoming gray

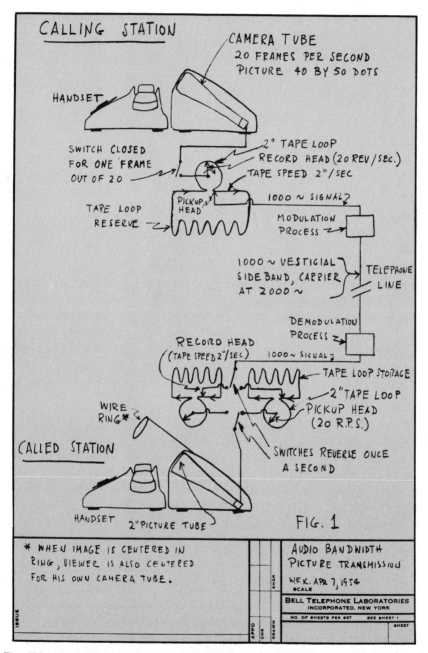

Fig. 176. A circuit diagram taken from a Bell memorandum of the author's showing how the tape device in Fig. 169 can provide narrow-band telephone television.

when the signal is weak, and remaining white for extremely weak signals. (It was this type of paper that was used in the original real time sound analyzer; Fig. 171 was made on this type of paper.) As the viewed person moved about, each picture would change. Only one "picture viewing" window would exist, and each picture would be presented for 2 seconds. Appendix 101 shows Ralph Miller (who, along with Ira Cole, developed this technique) holding a series of repeated (permanent) pictures made by this process.

The first full announcement of the experimental equipment occurred when a special evening session of the West Coast Institute of Radio Engineers was organized for the purpose of presenting a group of papers describing the Picturephone experiments.

THE ANNOUNCEMENT TO THE PRESS

The Bell announcement, on August 23, 1956, of the first Picturephone understandably evoked wide coverage by the press. But the news of the experiments apparently leaked out much earlier, resulting in a Newark *Sunday News* report on February 26, 1956 (Appendix 102); it included a picture of Bell Laboratories President Mervin J. Kelly and was titled "TV Phone Termed Feasible," with a subtitle "Cost Chief Hurdle, Says Lab Head."

As mentioned above, Bell's August 23 press release was timed to coincide with the presentation the night before of papers that I and my group had written describing the system's technical features at the 1956 "Wescon" Show and Convention. Appendix 103 shows the cover of the journal *Electronic Daily*, picturing three of my associates, Ralph L. Miller, Harold L. Barney, and Floyd K. Becker. The official August 23 Bell press release included the following:

> A telephone that transmits pictures along with sound has taken a big step towards commercial feasibility, Bell Telephone Laboratories announced last night (Aug. 22).
>
> Scientists and engineers at the Bell System's research and development organization have used an experimental "Picturephone" system to transmit recognizable pictures over short and long distances, even as far as from New York to Los Angeles.
>
> A technical description of this system was given last night at a joint meeting in Los Angeles of the Institute of Radio Engineers and the West Coast Electronic Manufacturers' Association.

Experimental pictures vary in size from one by one and a half inches to two by three inches, and are viewed from one to two feet away. Unlike television, a new picture is displayed every two seconds. It has good black-and-white contrast and the person at the other end of the line is recognizable. Head and shoulders can be seen and facial expressions are readily apparent.

The Picturephone described last night is the first system of its kind to use a pair of ordinary telephone wires. This is what gives it promise of being commercially feasible. Only one other line, consisting of a pair of wires like the regular telephone line, would be installed on the customer's premises to carry the picture.

It will be possible for the caller's picture to be "dialed" like an ordinary telephone call, provided the switch on the picture equipment is turned on at both ends of the line. If the switches are off, the telephone call will be completed without pictures. The picture can also be turned on after a conservation is under way. It would be impossible for a customer to be seen by the caller unless he flipped on the switch.

The Picturephone system differs from television in this way: TV sends 30 pictures a second and uses high frequencies which require relatively expensive coaxial cable and microwave radio relay systems, and provide an extremely detailed picture of a large scene. The Picturephone sends a smaller and less detailed picture every two seconds. The fact that this image can be transmitted over standard low-frequency telephone channels makes it possible to use ordinary telephone wires, like those running to the homes of telephone customers.

The picture equipment is still undergoing development and evaluation and is not ready for manufacturers or commercial use. There are several experimental units, different in size and appearance, but they operate on the same basic principles. Bell Laboratories engineers believe that a system based on these principles can eventually overcome the cost barrier.

The Picturephone, after further development, could be offered as a separate, optional telephone service. The unit would be made much more compact than the present experimental apparatus.

One of the experimental Picturephones as it has been operated in the laboratory uses a miniature television camera mounted on a desk. Another experimental unit incorporates the camera into the same unit as the screen. Instead of the bright lights required on a TV stage, the Picturephone will require only the natural daylight ordinarily available in a bright room, or moderate amounts of artificial light.

Present picture viewing apparatus is about the size of a small suitcase. Bell engineers expect to reduce the size drastically. Associated picture transmitting and receiving equipment must also undergo further development and design work.

A Picturephone caller can check his position in front of the camera by looking into a visual guide. As long as he is within the bounds of the guide, he is in the proper position to be photographed.

The Picturephone system was devised by Winston E. Kock, Floyd K. Becker, R. L. Miller, and others. Development work was directed by Kock. Miller and Becker are conducting new tests to explore the possibilities of various electronic gear to make the picture apparatus smaller than the present experimental setup.

ADDITIONAL BACKGROUND INFORMATION

Additional "background information" was also included in the Bell release; it provided detailed technical information. Thus the following description of the paper-record technique was included:

> Still another receiving system which is being investigated at Bell Laboratories is the use of a facsimile-type of recorder which prints the pictures on a small strip of paper, one every two seconds. Brightness is then limited only by the illumination on the picture once it appears on the strip, and a prism arrangement can be employed for viewing purposes so that to the viewer the image remains stationary for the full two seconds even though the strip of paper is moving. This system has the added advantage that the pictures are available as a permanent record for future reference.

PRESS REACTION

The concept of an inexpensive picture attachment to the telephone was apparently thought by the press to be of great interest to the public, and the coverage of the Bell announcement was extensive.

Thus arrangements were made for me to be interviewed (Appendix 104) on the "Eyewitness" program of KTLA-TV. In this interview, the early experimental movie of the secretary and me was shown, and numerous questions were put to me by the interviewer.

The *Wall Street Journal* featured the story on its front page (Appendix 105); the *Los Angeles Herald Express* carried its reports on its page 1 of Section D (Appendix 106), and the weekly *Science News Letter* carried a picture on the cover of its August 25 issue. The *American Weekly* referred, of course, to its 1906 article (Appendix 100) and stressed particularly the feature that "You can even carry a conversation with someone, and then hang up and pull out of the receiver a picture of the person who was on the other end of the line."

Naturally, humorous writers made much of the announcement.

The following example, entitled "The Face-to-Face Phone Call," is from H. I. Philips's syndicated column "The Once Over":

> The Bell Laboratories have developed an invention which will enable telephone users to see each other while they gab. It is an application of TV to phone service and is already being used in some areas. You can dial a face just as you dial a number. Since people frequently look their worst when they make a phone call, there can be embarrassments. Unless the telephone company distributes makeup kits as part of the communications service. Imagine the blow to romance when the girl, who is still wearing curl-pins, talks with the boyfriend who hasn't had time to shave yet! And give a thought to the husband, obviously a cocktail party victim, convincing his wife over a "Look-a-phone" that he is tied up in an important business conference.
>
> We all wonder about getting real conviction into it when a worker phones the boss to say he is too sick to come in to work, unless he never looks healthy anyhow. In a case of crossed wires, can the missus in one of those morning phone calls get part of Mrs. Fitchley's face and part of Mrs. Abernathy's? If she dials into Mrs. Clabby and gets six or seven other faces, will it mean a busy wire?

Also in a humorous vein, I received a copy of a German newspaper report of the experiments in which the sender had substituted for the secretary's face a picture of the well-known German professor Erwin Meyer, whom I had known since earning my doctorate at Berlin in 1934 (Appendix 107).

But perhaps the most satisfying to the Bell Picturephone group was the following editorial (particularly the last sentence) in the Sunday *New York Times:*

> SEEING AS WE TELEPHONE
>
> Why is it not possible to see the person to whom we talk over the telephone? The answer is that it is possible. In fact, the Bell Telephone Laboratories have been experimenting along these lines since 1927. The technical complications are formidable, and complications always mean heavy expense to the telephone company as well as to the subscriber. Now Bell engineers have presented at Los Angeles their latest effort in making it possible to see as we telephone. Despite its size and its clumsiness (it is still as big as a suitcase) the apparatus described at Los Angeles marks a real advance in communication engineering and this for the reason that only a pair of ordinary telephone wires is used—a marked simplification of the technical efforts made in the past.
>
> Let it not be supposed that what we have is a combination of television and telephoning. Pictures—what Broadway calls "stills"—are transmitted

as the subscribers talk to each other. Unlike television, a new picture is displayed every two seconds. What should appeal to future telephone users is the possibility of dialing a caller's picture just as we dial an ordinary telephone call, provided a switch at both ends of the line has been turned on. If he so wishes a subscriber may remain as invisible as he is now.

What's the use of seeing over a telephone as we talk? Public utilities have a way of finding unexpected uses. It is the public that develops the utilities. So will it be when we shall see one another as we talk over the telephone. "Let me see it," the housewife of the future may say as she telephones to the butcher for a roast chicken, whereupon he will obligingly hold up the bird before the camera. As for the business man, trust him to find a real need for seeing the man with whom he is talking. Facial expressions may count for much in the future. At any rate, the experiment described in Los Angeles deserves to rank historically with Alexander Graham Bell's famous request to his assistant Watson to "come here" when the telephone first talked.

THE MOVE TO HIGHER-QUALITY PICTURES

These first low-cost experimental systems were considered by most viewers to be reasonably adequate, and the requirement of only one additional phone circuit per home to provide this admittedly rather modest picture-viewing capability was a very important factor (as the above news reports indicated rather decisively). Bell, however, believing that the television-viewing public had become so accustomed to much higher-detail, continuous pictures on their television sets, decided to abandon the low-cost version and soon developed a Picturephone system which more closely matched broadcast TV standards. It provided continuous pictures which had detail (resolution) at least approaching that of TV broadcasts (it used a 1-megahertz channel as compared to TV's 4 1/2 megahertz). As noted in a 1969 Bell publication, the move required extensive modification of various parts of circuits which were designed for 4 kilohertz, not 1000 kilohertz. Such parts included crossbar switches, the network itself (insertion loss and "crosstalk"), and relays, which introduced serious noise in the picture signal.*

As we noted under the section "Transmission Cost," the new system obviously had to require the user to pay a much higher service charge than the charge for one 4-kilohertz telephone line, but Bell

---

* "The evolution of the Picturephone," *Bell Laboratories Record* (Special Issue), p. 160 (May/June 1969).

Fig. 177. Limited commercial Picturephone service between public locations in three cities—New York, Chicago, and Washington, D.C.—began on June 25, 1964. The service was inaugurated with a call from Mrs. Lyndon B. Johnson in Washington to Bell Laboratories scientist Dr. Elizabeth A. Wood, at the Picturephone Center in Grand Central Terminal, New York. Robert F. Wagner, then mayor of New York, is seated at right.

assumed that many would find the picture addition fully worth the much higher cost.

THE FIRST COMMERCIAL SERVICE

Eight years after the first announcement, on June 25, 1964, limited commercial Picturephone service between public locations in three cities—New York, Chicago, and Washington, D.C.—was inaugurated.* Figure 177 shows the inauguration event, which included the wife of President Lyndon Johnson, New York Mayor Robert Wagner, and Bell scientist Betty Wood, author of the well-received book *Science from the Air*. In 1968, an interesting experiment was conducted involving the use of the picture feature by deaf persons in "conversations" between New York and Washington.† Greetings, jokes, stories, and

* *Ibid.*
† *195 NEWS*, p. 2 (January 3, 1969).

gossip were exchanged by sign language and lip reading in this experiment conducted by New York University under a grant from the U.S. Department of Health, Education, and Welfare.

The 1969 *Bell Laboratories Record* special issue referred to earlier stated, in its introductory editorial, that "The Bell System is about to add Picturephone service to the many services it now offers customers" and that "commercial service is scheduled for mid-1970." However, in 1972, newspapers reported "the demise of the Picturephone," although, as noted, the "demise" was not of "the video telephone itself but only of the marketing of the Washington, D.C., services that was actually shelved."*

In 1973 an article noted that

> Picturephone service is almost nonexistent outside of Illinois, where Chicago has been a system testbed. Pittsburgh and Washington, D.C., both offer the service, yet Pittsburgh has but three customers; Washington, none. Illinois Bell currently lists 109 Chicago customers using 473 Picturephone sets—it's garnered this mini-universe by charging an admittedly promotional $86.50 per month, about half the $170 charged in Washington and $160 in Pittsburgh. Installation, extension, and per-minute tariffs are similarly sliced.
>
> About half the Chicago customers, including banks, law firms, architects, several advertising and printing firms, and almost a dozen hospitals, subscribe to *internal intercom* service only.†

Another article, under the paragraph title "A Disappointing Response," noted that "by the summer of 1972 there were only 56 non-A.T. and T. subscribers in the Chicago area."‡

At the price of $170 per month, the average person (who might enjoy quite often seeing low-detail pictures of relatives and friends) would quite likely consider it a "dubious assumption" that—at that cost—"face-to-face conversation would be the predominant use for the service." So Bell's most recent trend is to increase the bandwidth (and therefore the cost to the user) to the full 525-line (4.5-megahertz bandwidth,§ to provide further possibilities than just face-to-face conversations.

---

* Howard Falk, "Picturephone and beyond," *IEEE Spectrum,* p. 45 (November 1973).
† L. J. Hardeman, "Picturephone to change its image," *Electronics,* p. 75 (September 13, 1973).
‡ Falk, *op. cit.*
§ Hardeman, *op. cit.*

THE FUTURE

Someday, perhaps, Bell will offer *two* Picturephone versions, an inexpensive one aimed at the average homeowner and a second, full TV band one aimed at specialized users, capable of providing color pictures,* extremely high detail (by slow-scan facsimile transmission), interactive computer graphics, and, who knows, possibly even three-dimensional (holographic) images! The press coverage of the first models argues that the *first* of these two should be today's version, with the second being the better version when bandwidth costs (possibly through extensive use of domestic communications satellites) present a much smaller problem.

So, we have seen that early sound analysis procedures provided, as a spinoff, a way for telephone television at least to get its foot in the door, calling the attention of the public to Picturephone possibilities, through a low-definition, 2-second-repitition picture system. And soon we may be heading toward the day when a Picturephone will be in every home.

* Evolution of the Picturephone," p. 187.

PERORATION

Engineering science and technology have served mankind faithfully and well, have dramatically extended the average lifetime, shortened geographic distances, increased industrial productivity, and contributed significantly to the cause of freedom. If science and technology were to founder or stagnate, many of our hopes would collapse.

In 1971 a *Wall Street Journal* editorial had these words: "History has shown that the benefits of scientific discovery outweigh its risks. To fear scientific inquiry is to fear new ideas. And that, along with a susceptibility to hysterical thinking, could be a route to a new dark age."

We close this book with the observation that those who will enter the field of engineering are in large part those who possess, from birth, an aptitude for technology. They are, in effect, born to their calling. Some of their responsibilities are suggested in the following poem that reflects the fears of those who resent today's strong emphasis on technology. They remember the two nuclear holocausts, the intercontinental missiles, and the efforts of engineering scientists to expand their community. Its title, "The Technikers," simulates a German word for persons versed in science and technology:

> Holocaust fabricators
> Crafters of hurtling craft
> Influencers of world politics
> Universal users of universal broken languages
> Lobbiers for more, more, more of their ilk
> Avid, egocentric, glory-seeking searchers
> For breakthroughs in truth.
> The world no longer jeers and sneers
> It fears

The hordes who ravished twice
And now alone possess the secrets to survive.
Their myriad ideas,
Unlocking unknown doors to unknown awe-filled futures,
Float down on them like snowflakes
Seeking havens for their souls.

Their course was fully charted from the start
When in their youth they followed learning paths
Too devious for most.
Long hard exercise of abundant inborn ability
With constant craving for simple truthful truth
Led on inexorably to destined leadership.
This why they are
This why they should be
Holocaust fabricators
Crafters of hurtling craft
Influencers of world politics
Lobbiers for more, more, more of their ilk
Avid, egocentric, glory-seeking searchers
For breakthroughs in truth.

APPENDIX

LIST OF PAPERS PRESENTED AT THE .

URSI GENERAL ASSEMBLY - PARIS - 1946

Commission I

No. of document	Origin	Title and Author
9	France	Appareil de mesure de rayonnement sur ondes métriques et décamétriques par F. CARBENAY.
11	France	Etalons primaires et generateurs industriels pour les faibles tensions en haute fréquence par J. LOEB.
14	France	Procédés de comparaison rapide et précise des fréquences par B. DECAUX.
19	Gde Gretagne	Report on Radio Frequency Standards and Measurements in the United Kingdom.
45	Etats-Unis	A standard field-intensity recorder for receiving antenna calibration by V. C. PINEO
46	Etats-Unis	Notes for the design of a mutual inductance type attenuator, by J. J. FREEMAN
64	Etats-Unis	Microwave antenna measurements by C. C. CUTLER, A.P. KING and W. E. KOCK
72	Etats-Unis	Comparison of predictions of maximum usable frequency and lowest useful high frequency usage on radio communication circuits in 1944 by T. M. GAUTIER.
77	Etats-Unis	Oscilloscopic measurements on electrical transients in the millimicrosecond region by D. V. WINTER

Commission II

2	France	Sur quelques relations entre la lumière du ciel nocturne et les régions ionisées de l'atmosphère par A. et E. VASSY
3	France	L'influence des perturbations magnétiques sur la vitesse apparente de propagation des ondes courtes par N. STOYKO.
5	France	Rapport sur les perturbations ionosphériques par R. JOUAUST.

(over)

1. *First page of the program of a science meeting in Paris.*

FUTURE RELEASE: Tuesday morning, Feb. 1   1/28/49

GERMANIUM METAL REPLACING VACUUM
TUBES IN ELECTRIC TRANSISTORS

By SCIENCE SERVICE

NEW YORK, Jan. 31 --- A tiny bit of germanium metal may soon be used instead of a vacuum tube in radio receivers and other instruments where simple amplifying is required. In telephony and electronics in general the germanium device has great possibilities, the American Institute of Electrical Engineers was told here today by scientists of the Bell Telephone Laboratories. They revealed an improved germanium amplifier.

The device which utilizes this semi-conductor metal is called a transistor. Basic work on the transistor, which was announced in July, 1948, was carried on at the Bell Laboratories by Drs. John Bardeen and Walter H. Brattain. Later developments were described here today by Dr. Winston E. Kock and R. L. Wallace, Jr. All four men are members of the laboratory's technical staff.

These transistors, the meeting was told, are about the size of medicine capsules but can perform most of the key jobs now done

2. News release on the transistor.

THE WOODWARD PIANO DEPARTMENT

REQUESTS THE HONOR OF YOUR PRESENCE

AT A

PIANOFORTE RECITAL

PRESENTING

WINSTON KOCK

ASSISTED BY MISS ANNETTE FILLMORE

WOODWARD AUDITORIUM

SATURDAY EVENING   MAY 15, 1926

AT EIGHT O'CLOCK

3. Piano program, senior high school year.

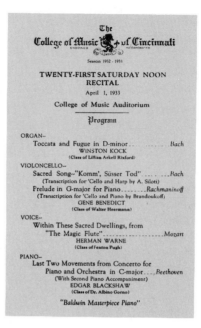

4. Organ program.

5. University of Berlin notification.

UNITED STATES PATENT OFFICE.

HENRY EDWARD KOCK, OF CINCINNATI, OHIO.

BUILDING CONSTRUCTION WITH REFERENCE TO COVERING MATERIAL.

No. 888,825. Specification of Letters Patent. Patented May 26, 1908.

Application filed January 2, 1908. Serial No. 408,963.

To all whom it may concern:

Be it known that I, HENRY EDWARD Kock, of Cincinnati, in the county of Hamilton and the State of Ohio, have invented 5 new and useful Improvements in Building Construction, of which the following is a full, clear, and exact description.

My invention relates to improvements in building construction with reference to 10 shingles, tiles or plates as a covering material making the structure composed of these more durable and lasting.

The invention consists in the novel features as hereinafter particularly described 15 and defined in the claims.

Reference is to be had to the accompanying drawing forming a part of the specification in which the Figure A shows a front view of such a shingle, plate or tile with the other 20 figures, B, C, D, E and F, illustrating various designs and combinations utilizing the same principle, however, in construction.

Shingles or tiles are usually constructed with a square end or having their longest 25 longitudinal dimension at or about the median line. In laying them as a covering for roofs, walls, etc., they are so placed next and above each other that one covers the seam formed by the abutment of the lower two. 30 In square ended shingles the water or drainage runs down to the square end and by capillary attraction is drawn under this end where it remains and rots the shingle. To obviate this shingles and tiles have been con-35 structed having their longest dimension at the median line so that the water will collect at one point and then run off. This however conducts the water into the crevice formed by the lower two shingles and causes these to rot. 40

In my invention the shingles, plates or tiles are so constructed that the greatest longitudinal dimension is at one or the other lateral edge as shown in Fig. A. In laying 45 shingles, plates or tiles of the above construction, in the usual manner, the water or drainage runs along each shingle, plate or tile until it reaches its lowest point, which then conducts it to the middle of the underlying 50 shingle or tile and not into the seam of the lower two thus preventing decay or corrosion. In like manner as shown in Figs. E and F each succeeding shingle or tile repeats the above process of drainage conduction there-55 by preventing the collection or accumulation of the drainage under or between the shingles or tiles preventing decay and thus prolonging the endurance of such covering material.

In Figs. B, C and D are shown different 60 patterns of shingles, tiles or plates for covering material to show the artistic effect to be achieved still utilizing the principle illustrated in Fig. A, while in Figs. E and F are illustrated shingles or tiles in position to 65 show how the principle of conduction is applied.

Having thus described my invention, I claim as new and desire to secure by Letters Patent:— 70

A shingle, tile or plate of suitable material having its lower margin cut to an acute angle with one only of its lateral edges.

HENRY EDWARD KOCK.

Witnesses:

AUGUST H. BROLEP,

ROBERT A. LE BLOND.

6. *Patent of Henry E. Kock. Other patents included No. 1.205,223 (1916).*

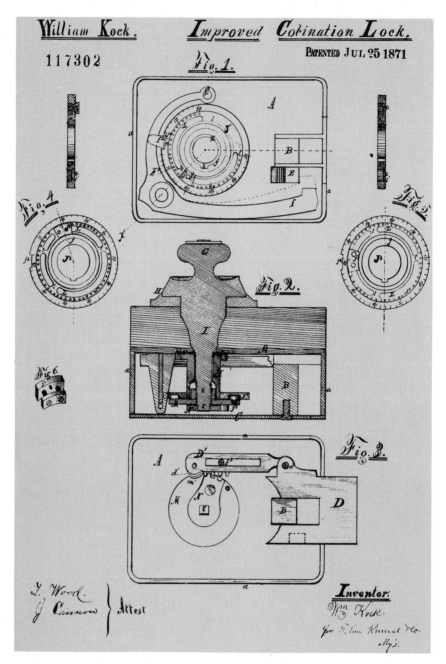

7. *Patent of William Kock. Other patents included Nos. 104,961 (1870), 146,191 (1874), and 586,930 (1887).*

Patented Dec. 6, 1938

2,139,023

UNITED STATES PATENT OFFICE

2,139,023

ELECTRICAL GENERATION OF MUSICAL TONES

Winston E. Kock, Cincinnati, Ohio, assignor to The Baldwin Company, Cincinnati, Ohio

Application August 23, 1935, Serial No. 37,547

16 Claims. (Cl. 84—1)

My invention relates to improvements in musical instruments of the type employing electrical oscillators of audible frequency for the production of tone.

It is found desirable in such electrical musical instruments to be able to produce many types of timbres or tone colors with one oscillator or with one set of oscillators in order to permit variations in registration. In the present invention I present several methods of improving the performance of an electrical musical instrument by providing distinct and novel means for varying its tone quality and tonal effects.

More specifically, I propose to set forth the manner of combining certain circuits to form tones resembling the different stops of a pipe organ or that can be combined in such a manner as to imitate some of the orchestral timbres.

One such method is the introduction of so-called "formants" or damped oscillations. It is known, for example, that the characteristic tone color of brass instruments, such as the trumpet or trombone, is imparted by the bell of the instrument. The pulsations of air being expelled from the trumpet set the flared bell into damped vibrations and these are heard superimposed upon the pulsating note issuing from the instrument. Excessive damping of these already slightly damped oscillations by means of a mute

several contacts on one key so that in addition to the basic tone being heard along with its electrically excited formant, the oscillators generating those tones which would correspond to the desired partials are connected to the amplifier and impart their timbre to the final tone.

Referring now to the drawings, Fig. 1 is a diagram of a circuit for producing formants of any desired damping.

Fig. 2 is a diagram of a heavily damped wave form.

Fig. 3 is a weakly damped wave form.

Fig. 4 is a periodic recurrence of damped oscillations.

Fig. 5 shows a formant circuit excited by an inductive glow discharge oscillator.

Fig. 6 shows the impulses of the voltage of the inductive glow discharge oscillator which act upon the formant circuit.

Fig. 7 shows the wave form of the output voltage and indicates how the formants are generated by the excitation impulses of Fig. 6.

Fig. 8 shows a formant circuit arranged for the production of low formants and excited by an inductive glow discharge oscillator.

Fig. 9 shows a formant circuit excited by a plurality of oscillators.

Figs. 10, 12 and 13 are modifications of formant circuits.

8. Patent describing the electronic organ formant concept.

Program includes Baldwin patent

A Baldwin organ patent is included in a Bicentennial slide program showing the workings of the patent system in the United States from a historical standpoint. Developed by the American Patent Law Association, the program was recently presented for the first time at the organization's fall meeting.

Patent 2,233,948, received by Dr. Winston E. Kock and assigned to Baldwin Piano Company, is the patent included in the program. This pre-World War II patent was the basis for Baldwin's entry into the organ field in 1946. The patent discloses the invention of a new general type of electronic organ now known as a "formant" organ.

Within a few years the formant organ dominated the previous harmonic-synthesis organ competition in the church field. When this patent and other Kock patents began to expire in the late 1950's, several other producers of electronic organs operating on

different principles abandoned these principles and began to produce formant organs.

At Baldwin Dr. Kock's research association and sometimes co-inventor was John F. Jordan, a former Baldwin vice president and still a company director. Dr. Kock later was associated with Bell Telephone Laboratories and Bendix.

9. U.S. Patent Association formant organ recognition.

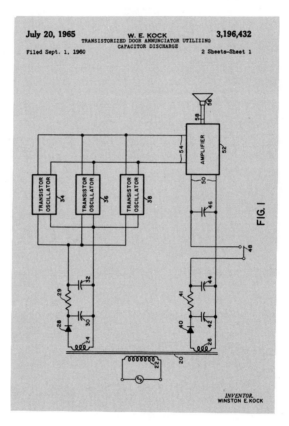

July 20, 1965   W. E. KOCK   3,196,432
                 TRANSISTORIZED DOOR ANNUNCIATOR UTILIZING
                          CAPACITOR DISCHARGE
Filed Sept. 1, 1960   2 Sheets—Sheet 1

FIG. I

INVENTOR.
WINSTON E. KOCK

10. First transistor doorbell patent.

11. A. M. Prokhorov, one of the three joint recipients of the Nobel Prize in Physics in 1964 for contributions to the field of masers, entertaining the author at his laboratory at the Lebedev Institute in Moscow.

12. Academician and Nobel Laureate Nikolai Basov (left) discusses his latest laser fusion experiments with the author in his office at the Lebedev Institute in Moscow. Basov shared, with A. M. Prokhorov and Charles Townes, the Nobel Prize for the discovery of stimulated emission (the maser and the laser).

INTERCOLLEGIATE CHESS CHAMPIONS OF OHIO. Members of the chess team, University of Cincinnati, to be seen in this group with their trophies, left to right, are Meyer Zeligs, Robert Teegarden, Winston Kock (Captain) and Ernst Theimer. The team won over all rivals in a tournament at Dayton.

13. A picture appearing in the Cincinnati Enquirer.

14. Winners of the 1930 chess problem composing contest.

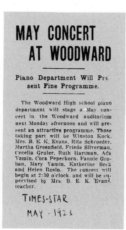

MAY CONCERT AT WOODWARD

Piano Department Will Present Fine Programme.

The Woodward High school piano department will stage a May concert in the Woodward auditorium next Monday afternoon and will present an attractive programme. Those taking part will be Winston Kock, Mrs. B. E. K. Evans, Rita Schroeder, Martha Greenfield, Frieda Silverman, Cecelia Gruler, Ruth Hartman, Ada Yamin, Cora Peperkorn, Fannie Groban, Mary Yamin, Katherine Beck and Helen Rosin. The concert will begin at 2:30 o'clock and will be supervised by Mrs. B. E. K. Evans, teacher.

TIMES-STAR
MAY - 1923

15. Newspaper account of a piano concert in which the author participated (at age 13).

PIANOFORTE RECITAL

BY WINSTON KOCK
ASSISTED BY
MISS ANNETTE FILLMORE

PROGRAM

Second Hungarian Rhapsody . . Liszt

Hunting Song
Consolation } . . Mendelssohn
Spinning Song

Sonata . . . . . C. P. E. Bach
   Allegro deciso   Andante cantabile
         Allegro vivace

A group of songs   Miss Annette Fillmore
   When The Roses Bloom Reichardt
   Pierrot   Wintter Watts
   Sylvan   Ronald

If I were a Bird . . . . Henselt

Impromptu in A flat . . . Schubert

Egeria . . . . . E. R. Kroeger

Raindrop Prelude . . . Chopin

Peer Gynt Suite . . . . Grieg
   Morning
   Death of Ase
   Anitra's Dance
   In the Hall of the Mountain King

16. Program of the high school piano recital.

17. *A widely used (copyrighted) church music composition.*

Newcomer Reaches Success In Musical ComedyProduction

Winston Kock, Engineer Writes Over Half Of Musical Numbers

"Comedy Work Entertaining And Educational," Says Talented Musician

A newcomer crashed the Musical Comedy gate and in his first season with the Fresh Painters has made a name for himself. Winston Kock, a Junior in Electrical Engineering, has written over half the music for "G'wan And Kiss Her."

Kock is a graduate of Hughes High of the class of '26, and a brother of Fred Kock, well-known Applied Arts alumnus. He is a member of Tau Beta Pi, honorary engineering fraternity, and Eta Kappa Nu, electrical engineering honorary. His success in the composing line is due to an extentive musical training. He has had courses in composition and harmony at the College of Music and studied piano under Mrs. Evans. At present he is in the class of Mrs. Ricksford of the College of Music, studying the organ, an instrument which he prefers to the piano. At Hughes High school Kock also entered into various musical activities.

Kock has written nine compositions for this year's production, many of which are to be used as repeats in the finales. The most outstanding of his contributions to the musical score are "The More I Think of You," a lead solo sung by Bill Hudson and Peg Terry, and a sequidillo, a Spanish dance in classic style.

18. *Newspaper account of musical comedy composer activity in 1931.*

19. *The first page of a Spanish dance number which was part of a 1931 University of Cincinnati amateur musical comedy.*

STUDENTS COMPOSING SCORES FOR COMEDY

WINSTON KOCK

As a member of the Fresh Painters' music committee last year, Winston Kock attracted much favorable attention with the twelve original scores he contributed to "Sittin' Pretty." Among his compositions many of the favorites of the show are to be found.

Kock is serving as co-chairman of music this year and also contributing six original selections. His compositions show a light and tuneful quality, particularly adaptable to the fantasy "Tarts Are Trump."

Besides having written many compositions of a popular nature, Kock has composed some church music. He specializes in organ study, playing regularly at the Church of the Nativity in Price Hill. He is a student at the College of Music.

Kock is a senior in electrical engineering. He has recently been elected to Sigma Xi, honorary research fraternity. Kock holds offices in Tau Beta Pi, honorary engineering scholastic fraternity, and Eta Kappa Nu, honorary electrical engineering fraternity.

Kock is president of the Student branch of the American Institution of Electrical Engineers. He is president of the Chess Club and captain of the Chess Team.

20. *Newspaper account of composing activities.*

◀ MUSICAL NUMBERS ▶

ACT I

1. Opening Chorus (Tarts are Trump)..Dancing Girls, Show Girls, Men
   Music by Winston Kock—Lyric by Willa Busch, Dan Tobin, Winston Kock,
   Fred Scull

1A. Surest Way to the King's Heart... Queen
   Music by Winston Kock—Lyric by Willa Busch

2. Father's Side of the House.....The Jack, Cook, Joker, Dancing Girls
   Music by Phyllis Kasle—Lyric by Dale Richardson

3. Man I Want...................................Princess and Retinue
   Specialty..Elsa Trefzger
   Music by Clara L. Zinke—Lyric by Willa Busch

4. It's Easy to Fall in Love...........The Jack, Princess, Show Girls, Men
   Music by Winston Kock—Lyric by Willa Busch

5. It's a Woman Every Time.........................Ace, Parliament
   Music by Winston Kock—Lyric by Willa Busch

6. Oh, How We Hate that Guy................Joker, Cook, Dancing Girls
   Music by Winston Kock—Lyric by Eleanor Brill

7. Finale .... Ensemble

21. Act I musical numbers.

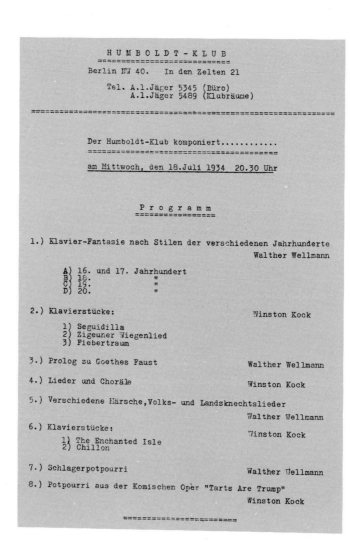

H U M B O L D T - K L U B
==================================

Berlin NW 40.   In den Zelten 21

Tel. A.1.Jäger 5345 (Büro)
    A.1.Jäger 5489 (Klubräume)

==

Der Humboldt-Klub komponiert...........
==

am Mittwoch, den 18.Juli 1934 20.30 Uhr

P r o g r a m m
=====================

1.) Klavier-Fantasie nach Stilen der verschiedenen Jahrhunderte
                                        Walther Wellmann

        A) 16. und 17. Jahrhundert
        B) 18.   "
        C) 19.   "
        D) 20.   "

2.) Klavierstücke: Winston Kock

        1) Seguidilla
        2) Zigeuner Wiegenlied
        3) Fiebertraum

3.) Prolog zu Goethes Faust   Walther Wellmann

4.) Lieder und Choräle   Winston Kock

5.) Verschiedene Märsche,Volks- und Landsknechtslieder
                                 Walther Wellmann

6.) Klavierstücke: Winston Kock
        1) The Enchanted Isle
        2) Chillon

7.) Schlagerpotpourri   Walther Wellmann

8.) Potpourri aus der Komischen Oper "Tarts Are Trump"
                                 Winston Kock

            =========================

22. *Berlin program of piano compositions.*

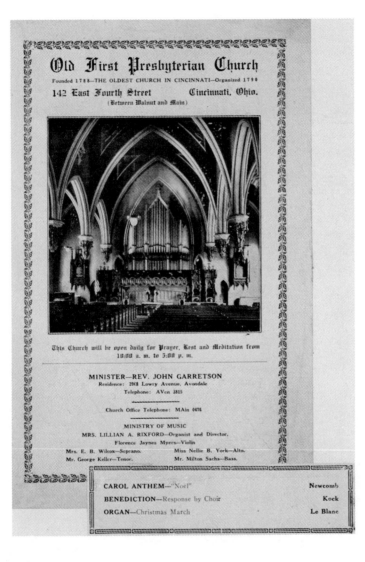

23. *A composition used in a large Cincinnati church.*

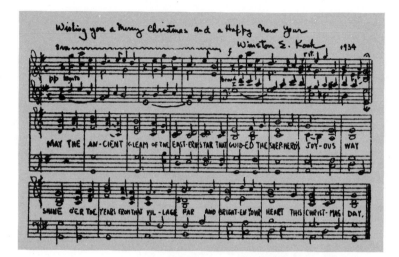

24. *The author's music is employed in his 1934 Christmas card.*

25. *Christmas card including the first son.*

26. *The 1949 Christmas card included the author's children, Winston Jr., Robert, and Kathleen.*

27. *The 1971 Christmas card featuring the first grandchild, Cristy.*

28. The 1975 Christmas card featuring the full Kock family.

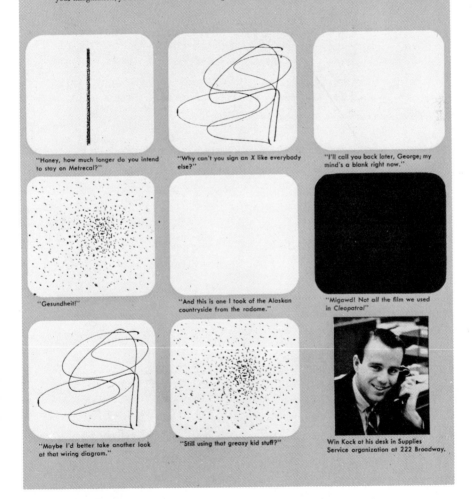

RIDICULES

Beauty, it is said, is in the eye of the beholder. And so is humor. If you look at the panels below from a literal point of view, you will see only scribbles, dots and solids. But if you let loose your imagination, you can see all sorts of amusing and topical situations portrayed—you can, that is, if you're Winston Kock of Supplies Service. Win has been dreaming up humorous captions for abstractions for some time. Below, a few that tickled our funnybones.

"Honey, how much longer do you intend to stay on Metrecal?"

"Why can't you sign an X like everybody else?"

"I'll call you back later, George; my mind's a blank right now."

"Gesundheit!"

"And this is one I took of the Alaskan countryside from the radome."

"Migawd! Not *all* the film we used in Cleopatra!"

"Maybe I'd better take another look at that wiring diagram."

"Still using that greasy kid stuff?"

Win Kock at his desk in Supplies Service organization at 222 Broadway.

29. *Cartoons dreamed up by the author's son.*

Clifton Public School
June 1915

Winston Koch

Promoted to
Second Grade
Excellent

30. Promotion card to second grade received at age 5.

UNIVERSITY OF CINCINNATI
COLLEGE OF ENGINEERING AND COMMERCE

Cincinnati, Ohio
June 30, 1926.

Mr. Winston Koch,
3280 Jefferson Ave.
Cincinnati, Ohio.

Dear Sir:

Your scholarship record was received. We note that
you are 16 years of age. We
regret to advise you that the
rule of the College of Engineering
is that a student under 17 is not
eligible for admission.

Very truly yours,

Dean, College of Engineering
and Commerce.

K

31. Letter indicating age requirements.

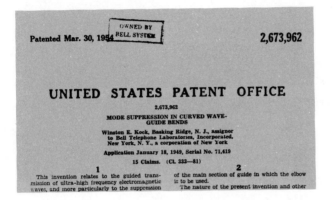

32. *A patent on a technique for suppressing unwanted modes in a circular waveguide.*

DATE Jan 27, 49 **131**

CASE No. 20564

The idea of using a wave guide as a means of changing the velocity of microwaves by varying their frequency and thus obtain a "velocity modulation" effect whereby the individual energy packets fed into a wave guide at different time can be made to catch up with one another by increasing the group velocity of the waves entering later, was considered:

Input
F.M. f_0 has group velocity v_0 √ Catcher point

$f_0 \rightarrow f_0 + \Delta f$ $f_0 + \Delta f$ has higher group velocity $v_0 + \Delta v$

$f_0 + \Delta f$ entered input at later time but due to higher group velocity it arrives at catcher point coincident with f_0, exhibits itself as higher voltage or higher energy at the catcher point than would have appeared there without F.M. If F.M. is periodic, the bunches arrive at catcher with the same periodicity and it may be possible to build up very high voltages (pulse peaks) at the modulating frequency. If the amount of velocity modulation is proportional to signal (amplitude) modulation at the catcher will be similar to the applied signal. The device may thus be used as a discriminator (F.M. to A.M.)

                                               W. C. Hahn 1-27-47

33. *A notebook entry describing wave bunching.*

BELL TELEPHONE LABORATORIES
INCORPORATED

COVER SHEET FOR TECHNICAL MEMORANDA

SUBJECT: Velocity Modulated Waves - Case 38138-3

COPIES TO:
1 —
2 —Case File
3 —RKP-RB-Dept. 1000 Files
4 —Central Files (4)
5 —R.K.Potter (with ltr. of)
6 —H.C.Hart (with copy of ltr. to RKP)

MM— 49-130-21
DATE June 6, 1949
AUTHORS W. E. Kock
         F. K. Harvey

Velocity Modulation

ABSTRACT

Experiments are reported upon in which a constant-amplitude, frequency-modulated signal is transformed into a pulse after passing through a transmission line in which the wave group velocity depends upon frequency. This "wave bunching" effect is similar to the bunching of electrons in a velocity modulated electron stream. Because of the wave nature of electrons, modification of the electron bunching analysis is indicated.

34. A memorandum describing wave bunching using a corrugated tube.

35. Discoverer of the SOFAR channel, Professor Maurice Ewing, stands in the Bell Laboratories quiet room. Holding a seashell to his ear, he confirms his conviction that the rushing noise normally heard when a shell is held to the ear is a resonance phenomenon, requiring an ambient noise for its creation. In exceedingly quiet surroundings, the seashell is devoid of sound.

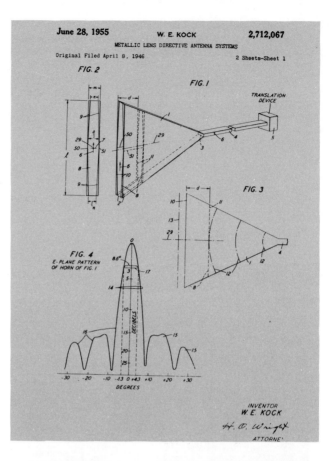

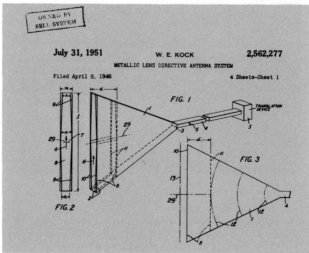

36. *The patent drawing of the author's first waveguide lens.*

SECRET

JUL 11 1946

CHIEF, SECURITY AND NOMENCLATURE SECTION
Headquarters, Services of Supply
Office of the Chief Signal Officer
Washington, D. C.

Dear Sir: Attention: Mr. G. E. Downer

      Enclosed herewith are the following Bell Telephone
Laboratories reports:

    MM-44-160-67   "Experiments with Metal Plate Lenses
                     for Microwaves" by W. E. Kock

    MM-44-160-100   "Wire Lens Antennas" by W. E. Kock

    MM-44-160-195   "Metal Plate Lens Design Considera-
                     tions" by W. E. Kock

    MM-45-160-23   "Metal Plate Lenses for Microwaves"
                     by W. E. Kock

These reports were requested in your letter of June 19, 1946
Ref: SIGTM-38.

      Even though this material may subsequently be
declassified, it is not released for publication.

Very truly yours,

ORIGINAL SIGNED BY
R. K. HONAMAN
DIRECTOR OF PUBLICATION

Att:
MM-44-160-67, 3/27/44
   44-160-100, 4/22/44
   44-160-195, 8/10/44
   45-160-23, 3/23/45

Copy to
Mr. W. E. Kock —— THIS COPY FOR

C O P Y
BELL TELEPHONE LABORATORIES, INC.
SECRET

K-7111(10-43)

37. World War II secrecy measures reduced information transfer.

Cincinnatian Paid Tribute

By Dr. J. O. Perrine, Assistant
Vice President, A. T. & T., in his
talk before Engineering Society.

Dr. Perrine, using small metallic lens, developed by Winston
E. Kock, Cincinnati scientist, to
reflect wave and light neon
bulb.

Dr. J. O. Perrine, Assistant Vice President, A. T. & T., recognized and emphasized the contribution made to radio by Winston E. Kock, Cincinnati scientist, now affiliated with Bell Laboratories, when he spoke at the May 16 meeting of the local Engineering Society.

The metallic lens, developed by Kock, concentrates radio micro-waves into a narrow beam where they can be controlled and thus increase their utility value. By preventing dissipation all over the air through the medium of this lens it would be possible to send 500 messages and two television broadcasts simultaneously over the concentrated radio beam. These lenses were used in testing the New York-Boston radio-telephone service installation.

Demonstration of this lens and other graphic illustrations featured Dr. Perrine's talk on "Radar and Microwaves". The speed of the microwave was reported at 1000 feet in one millionth of a second. Wave form and length was illustrated by the use of a nylon rope stretched between a motor driven vibrator and a wave reflector.

In speaking of radar and its accomplishments, Dr. Perrine told of the experience of the crew of the South Dakota. This battleship was the goal of 38 Jap planes sent out to destroy it, but through the medium of the radar equipment the crew was able to spot and knock out all of the enemy planes without damage to the ship.

38. *The outstanding Bell public relations scientist J. O. Perrine speaks about the micro-wave lens.*

39. *Assembly of a large strip lens using metal foil strips inserted in slots in plastic foam. The photo shows the author's colleague William Legg.*

Vol. 3, No. 83   May 27, 1948

SCIENCE TO-DAY

RADIO LENSES
ROYAL SOCIETY EXHIBITS
BONES AND MATURITY
IN BRIEF
SOLAR DISTURBANCES
BOOKS

Radio Lenses

A LENS, to most of us, means a piece of glass or quartz of which two opposite surfaces are close to spherical.

The latest type of lens designed for the "beaming," not of light, but of centimetre radio waves, has no such solidity. It consists instead of a pattern of metal spheres or flat discs, mounted on insulating material, and given an overall shape corresponding with the wanted lens form.

It is due to W. E. Kock of the Bell Telephone Laboratories. His aim was to produce a practical device. But the fact that his lens will work demonstrates on a vastly greater scale what happens in any normal optical lens—and it was, in fact, from this point of view that he approached the problem.

* * *

A lens, whether of glass or quartz, is not—as we now know —a continuous material. It is a collection of atoms, and these atoms are surrounded by electrons which can vibrate.

The network thus provided is finer, by a factor of several thousand, than the length of visible light waves. And the fact that any particular material is translucent means essentially that its atoms are capable both of absorbing energy and of re-radiating it at the same frequency as the light which is passing through it.

* * *

The further property of refraction, of which any lens makes practical use, is due to the slowing up of the advancing wave disturbance because of interference between the original light waves and those re-radiated.

The effect depends quantitatively, as might be expected, both on the number of atoms and the capacity of individual atoms to absorb and re-radiate.

Such, in outline, is the theory. Kock, looking at it, saw that for the longer waves of radio "a scaled-up version or model should also focus centimetre waves, equally scaled up in wavelength." His "radiations," correspond-

Edited by A. W. Haslett, M.A., from 104 Clifton Hill, London, N.W.8, for Weekly Science Newsletter, Ltd. Telephone: Maida Vale 5779. Subscription rates: 30s. a year (50 issues) or 15s. 6d. six months (25 issues)

40. *The British science journal* Weekly Science Newsletter *features a story on artificial dielectrics.*

PRESS RELEASE *from* Bell Telephone Laboratories
463 West Street, New York 14 · CHelsea 3-1000

Release: Friday A.M., February 13, 1948.

An entirely new type of metal lens for focussing
radio waves in radio relay systems is under development at
Bell Telephone Laboratories. Present plans call for the use
of the lens, theoretically capable of handling from 50 to
100 television channels or tens of thousands of simultaneous
telephone messages, in the proposed radio relay link which
the Bell System is planning between New York and Chicago.
The new lenses are based on the theories of light transmission
through atomic and molecular structures and use metallic
spheres, discs or strips in a scaled-up pattern similar to
the arrangement of atoms in a crystalline molecule. One
type of the lens, shown being assembled above, employs strips
of metal foil supported in a lens-like arrangement in polystyrene foam.

(see attached story)

A-5171

41. The Bell press release on the artificial dielectric lens.

42. A photo recording the historic opening of the first transcontinental microwave relay system. Executives of the Federal Communications Commission and AT&T participated.

43. The televised address of U.S. President Harry Truman, at the opening session of the 1951 Peace Treaty Conference, was carried across the country via the coast-to-coast microwave relay circuit.

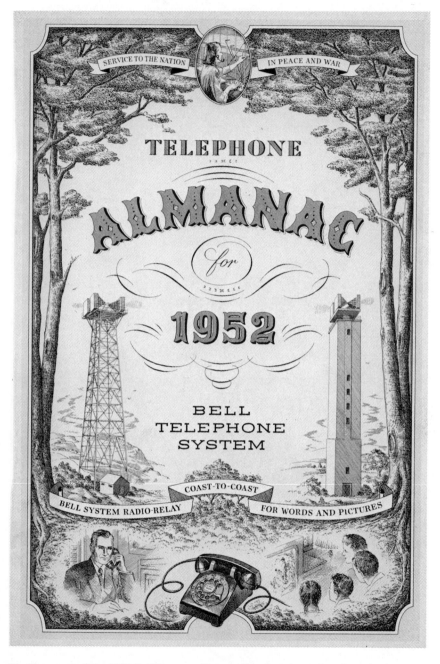

44. *The cover of the 1952 Bell Systems Almanac featured the radio relay circuit.*

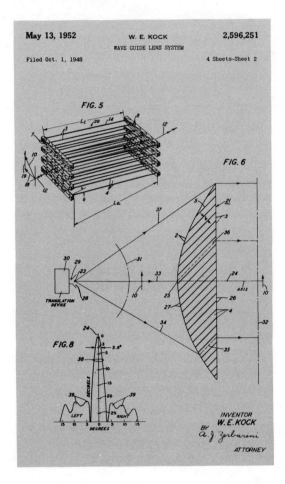

45. The patent application on the path length lens was filed in October 1948.

46. *The Bell Laboratories quite room at Murray Hill, New Jersey, during Edward R. Murrow's television broadcast.*

47. *The picture of F. K. Harvey and a sound wave lens featured on a journal cover.*

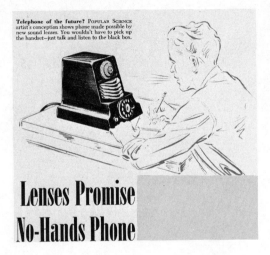

Telephone of the future? POPULAR SCIENCE artist's conception shows phone made possible by new sound lenses. You wouldn't have to pick up the handset—just talk and listen to the black box.

Lenses Promise No-Hands Phone

48. *A science writer's concept for using lenses with a "hands-free" telephone.*

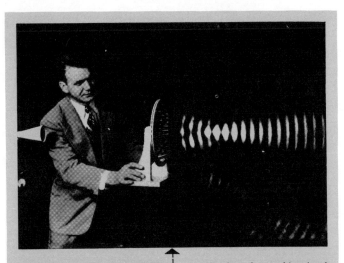

WAVE MAKING

—for better telephone service

Waves from the sound source at left are focused by the lens at center. In front of the lens, a moving arm (not shown) scans the wave field with a tiny microphone and neon lamp. The microphone picks up sound energy and sends it through amplifiers to the lamp. The lamp glows brightly where sound level is high, dims where it is low. This new technique pictures accurately the focusing effect of the lens. Similar lenses efficiently focus microwaves in radio relay transmission.

At Bell Telephone Laboratories, radio scientists devised their latest microwave lens by copying the molecular action of optical lenses in focusing light. The result was a radically new type of lens—the array of metal strips shown in the illustration. Giant metal strip lenses are used in the new microwave link for telephone and television between New York and Chicago.

The scientists went on to discover that the very same type of lens could also focus sound . . . thus help, too, in the study of sound radiation . . . another field of great importance to your telephone system.

The study of the basic laws of waves and vibrations is just another example of research which turns into practical telephone equipment at Bell Telephone Laboratories . . . helping to bring you high value for your telephone dollar.

BELL TELEPHONE LABORATORIES

Working continually to keep your telephone service one of today's greatest values

49. *This 1951 Bell advertisement suggested that sound wave focusing could be useful in telephone systems.*

A focus on better, low-cost telephone service

In the new microwave radio relay system between New York and Chicago, giant lenses shape and aim the wave energy as a searchlight aims a light beam.

Reasoning from the action of molecules in a glass lens which focuses light waves, Bell Laboratories scientists focus a broad band of microwaves by means of an array of metal strips. To support the strips these scientists embedded them in foam plastic which is rigid, light in weight, and virtually transparent to microwaves.

This unique lens receives waves from a wave guide at the back of the horn. As they pass across the strips, the waves are bent inward, or focused, to form a beam like a spotlight. A similar antenna at the next relay station receives the waves and directs them into a wave guide for transmission to amplifiers.

This new lens will help to carry still more television and telephone service over longer distances by microwaves. It's another example of the Bell Telephone Laboratories research which makes your telephone service grow bigger in value while the cost stays low.

Laboratory model of the new lens. A similar arrangement of metal strips is concealed in the foam plastic blocks in the large picture.

BELL TELEPHONE LABORATORIES

Working continually to keep your telephone service big in value and low in cost.

50. *A Bell advertisement featuring the strip lens for microwave use.*

NATIONAL ACADEMY OF SCIENCES
NATIONAL RESEARCH COUNCIL
2101 CONSTITUTION AVENUE, WASHINGTON 25, D. C.

Dr. Winston E. Kock
Bendix Aviation Corporation
Ann Arbor, Michigan

Dear Dr. Kock:

With the hectic pace of the last few weeks it has been difficult to keep intimately in touch with you and our other colleagues who participated in the recent National Academy of Sciences study for the Air Research and Development Command.

I was particularly sorry that it was not possible for us all to come together again before the briefing for the Air Force last week. This, as you undoubtedly know, marked the formal close of the activities conceived for this year, and was, I believe, very well received. Our written report has also been submitted to ARDC and I am hopeful that this will be circulated to all of you whose combined efforts made it possible to realize.

On behalf of Dr. Bronk and the National Academy of Sciences-National Research Council may I tell you how much we appreciate everything that you contributed to this year's endeavours. I, personally, hold in high regard the devotion and support that you gave to this project, and hope that it will be possible for us to work in close association again in the not-too-distant future.

Sincerely,

Theodore von Karman

Theodore von Karman

51. Dr. von Karman's letter regarding the NASA Air Research and Development Command's Wood's Hole Summer Study.

Three principal members of the National Academy of Sciences' Summer study group at Little Harbor Farm in Woods Hole discuss their research studies around an informal conference table. At the table, left to right, are Dr. Joseph Kaplan, professor of geophysics at the University of California and chairman of the U.S. committee for the International Geophysical Year; Dr. Edwin H. McMillan, Nobel Prize winner and member of the Atomic Energy Commission's scientific advisory committee, and Dr. H. Guyford Steever, dean of engineering at M.I.T. and vice-chairman of the Air Force's scientific advisory board.

52. *The Summer Study group of Item 51 questions a Russian announcement. (The author can be seen in the upper right.)*

53. *The announcement of Sputnik startles the world.*

54. *This poem expresses disappointment which was fairly widespread in the United States when the USSR announced the successful orbiting of the world's first manmade moon, Sputnik I. "Our golfer," referred to in the poem, is then President Eisenhower. Courtesy Associated Press.*

55. *Wernher von Braun's speedup of the placing of a U.S. "moon," Explorer I, into orbit generates widespread headlines.*

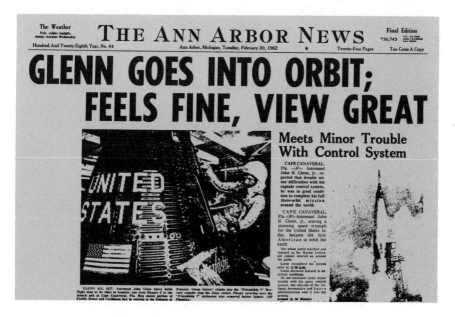

56. *Headlines mark the successful flight of John Glenn, the first U.S. astronaut to orbit the earth.*

57. *Astronaut Edward White took the first "walk" in space while his partner James McDivitt remained in the spacecraft.*

NEIL A. ARMSTRONG

58. An autographed photo of Neil Armstrong, the first man to set foot on the moon.

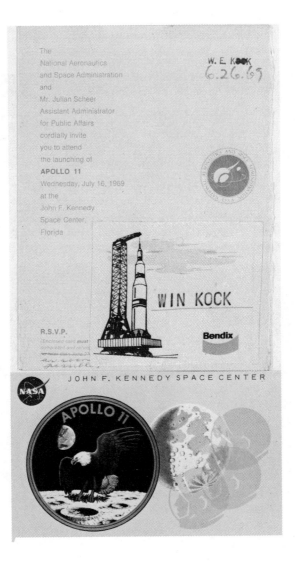

59. The invitation and the "badge" for the launch ceremonies for Apollo 11.

60. *The Bell System's Telstar provided the world's first transmission of television across the Atlantic.*

61. *Coverage of the United States, as sketched in 1969, for a "domestic" satellite. Courtesy Communications Satellite Corporation.*

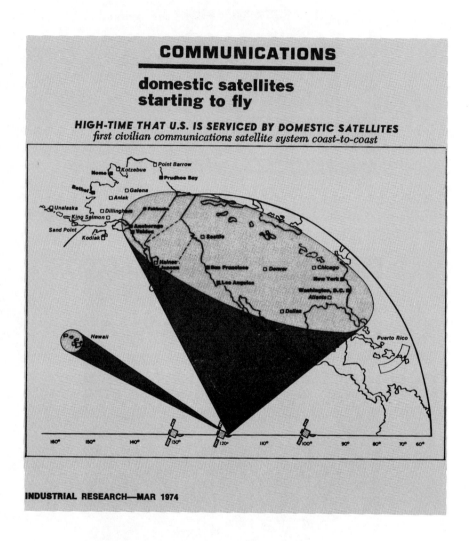

62. *The title conveys the impatience felt by many Americans regarding domestic satellites.*

ESTABLISHED 1862

The Baldwin Company

MANUFACTURERS

Baldwin · Acrosonic · Hamilton · Howard Pianos

CABLE ADDRESS
"BALDWINCO"

CINCINNATI 2

April 14, 1947

Mr. Winston E. Kook
Hill House
Red Hill Road
Middletown, New Jersey

Dear Sir:

      This Company is now engaged in the manufacturing
and sale of electronic organs. The instrument being manufactured embodies one or more of your sole or joint inventions as contemplated by agreement entered into by you with
this Company on June 28, 1938.

      First sales were made in January of this year.

The Baldwin Piano Company

REMITTANCE STATEMENT						DETACH BEFORE DEPOSITING
INVOICE DATE			M E M O	GROSS AMOUNT	DISCOUNT AMOUNT	NET AMOUNT
MO.	DAY	YR.				
4	12	56	ROYALTIES FIRST QUARTER.	2,971.00		2,971.00
NO. 018488			TOTALS			2,971.00

Enclosure

63. *Sales of the first few Baldwin electronic organs began during the first quarter of
1947 following the end of World War II. By 1956 the royalties had begun to be fairly
appreciable.*

64. The Ann Arbor home acquired by the author in late 1956.

65. An example of the wide coverage given to an industry press release.

From: Carl Byoir & Associates, Inc.   For Immediate Release
       800 Second Avenue, YUkon 6-6100
       New York, New York 10017

For: THE BENDIX CORPORATION

HIGH NAVY HONOR

AWARDED BENDIX'

DR. W. E. KOCK

WASHINGTON, D.C. -- The Navy's Distinguished Service Award was given re-

cently to Dr. W.E. Kock, vice president for research of The Bendix Corporation.

The award was made for Dr. Kock's services and technical contributions in

the field of anti-submarine warfare.

The presentation, made personally by Secretary of the Navy Paul H. Nitze,

was a highlight of a two-day meeting here of the National Industrial Security

Association's Anti-Submarine Warfare Committee headed by Dr. Kock.

The committee, comprising 600 members from 150 industrial concerns, present-

ed Secretary Nitze with a three-volume report covering industry's views on various

aspects of the anti-submarine warfare problem.

- 0 -

5/27/64

66. A press release announcing the highest possible award the Navy can present to a civilian; the citation included the words "In recognition of outstanding contribution to the Anti-Submarine Capability of the United States Navy."

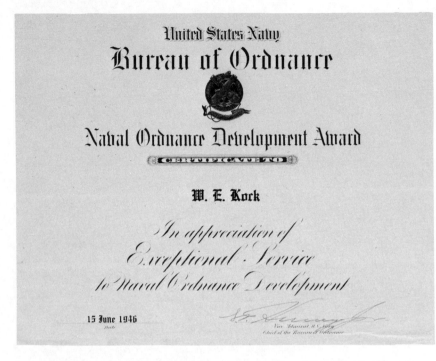

67. *A U.S. Navy award based on the Mark 13 radar discussed in Chapter 10.*

                    *A similar result has been seen in the   117. Kock Organ
        field of electronic organs. In the 1930's, Dr.
        Winston E. Kock invented the formant organ, an
        improvement over earlier electronic organs.
        Based upon Dr. Kock's patent the Baldwin Piano
        Company entered the organ field in 1946. ~~Quickly~~
        Gradually the formant organ dominated the previous harmonic
        synthesis organ competition *and upon expiration   118. Kock Patent
        of this patent, several other producers of elec-
        tronic organs adopted the formant principle, a
        tribute to Dr. Kock and to the proper operation
        of the patent system.

68. *A portion of the American Patent Law Association's Bicentennial essay.*

The device, about to be submerged, is an "underwater sound source". It transmits sound waves beneath the sea and is part of the research equipment developed by Bendix Research Laboratories Division, for use in the Bendix program to increase the capabilities of sonar.

UNDERWATER "EARS" TO HELP DEFEND
AGAINST ENEMY SUBMARINES

Perhaps no problem today requires a solution so urgently as the development of a defense against attack by enemy submarines.

Just as hundreds of long-range radar stations comprise our vast air attack warning system, *sonar* (a type of underwater radar which transmits sound waves instead of radio waves) is one of the undersea techniques that will be used to "listen" and warn of approaching submarine attack. But, unfortunately, sonar does not have the long-range capabilities of radar. Therefore, it is vital that the range of sonar be increased so that it can detect enemy submarines farther from our shores and permit us to take proper defensive action.

Because Bendix® is a pioneer in sonar research and development, and has supplied such equipment to our government for many years, it is one of the organizations selected to improve sonar performance and to help develop new techniques for defense against submarine attack.

This project is but one of many in a long and diverse list of Bendix developments in products that work beneath the sea. Another important Bendix anti-submarine device is "dunked" sonar. Here, helicopters lower the listening unit into the sea to spot enemy submarines. We also manufacture control-rod drive mech-

anisms for the nuclear reactors in nuclear-powered submarines and in the *USS Long Beach*, the Navy's first nuclear-powered cruiser.

The underwater "TV eye", which enabled the crew of the nuclear-powered submarine *Skate* to see the underside of the Polar ice pack, and to locate thin ice areas in which they could safely surface, was developed by Bendix. In addition, Bendix is a major manufacturer of depth recorders, of "brains" to guide torpedoes and undersea missiles, plus new types of hydraulic equipment for steering and diving operations and for controlling a score of other all-important functions on submarines.

A thousand products   *a million ideas*

AVIATION CORPORATION
Fisher Bldg., Detroit 2, Mich.

69. A Bendix advertisement featuring the antisubmarine research conducted by the author's Bendix group.

70. *A silver trophy won by the University of Cincinnati chess team when the author was captain.*

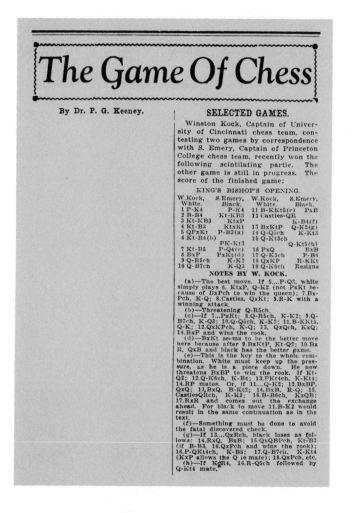

The Game Of Chess

By Dr. P. G. Keeney.

SELECTED GAMES.

Winston Kock, Captain of University of Cincinnati chess team, contesting two games by correspondence with S. Emery, Captain of Princeton College chess team, recently won the following scintilating partie. The other game is still in progress. The score of the finished game:

KING'S BISHOP'S OPENING.

W.Kock, White.	S.Emery, Black.	W.Kock, White.	S.Emery, Black.
1 P-K4	P-K4	11 B-KKt5(c)	PxB
2 B-B4	Kt-KB3	12 Castles-QR	
3 Kt-KB3	KtxP		K-B4(f)
4 Kt-B3	KtxKt	13 BxKtP	Q-K2(g)
5 QPxKt	P-B3(a)	14 Q-Q5ch	K-Kt3
6 Kt-R4(b)		15 Q-Kt3ch	
	P-K-Kt3		Q-Kt5(h)
7 Kt-B5	P-Q4(c)	16 PxQ	BxB
8 BxP	PxKt(d)	17 Q-K3ch	P-B4
9 Q-R5ch	K-K2	18 QxKP	R-KKt
10 Q-B7ch	K-Q3	19 Q-K6ch	Resigns

NOTES BY W. KOCK.

(a)—The best move. If 5....P-Q3, white simply plays 6. KtxP, Q-K2 (not PxKt because of BxPch to win the queen); 7.BxPch, K-Q; 8.Castles, QxKt; 9.R-K with a winning attack.

(b)—Threatening Q-R5ch.

(c)—If 7...PxKt; 8.Q-R5ch, K-K2; 9.Q-B7ch, K-Q3; 10.Q-Q5ch, K-K2; 11.B-KKt5, Q-K; 12.QxKPch, K-Q; 13. QxQch, KxQ; 14.BxP and wins the rook.

(d)—BxKt seems to be the better move here because after 9.BxKtP, Kt-Q2; 10.Bx R, QxB and black has the better game.

(e)—This is the key to the whole combination. White must keep up the pressure, as he is a piece down. He now threatens BxBP to win the rook. If Kt-Q2; 12.Q-K6ch, K-B4; 13.PKt4ch, K-Kt4; 14.RP mates. Or, if 11....Q-K2; 12.BxBP, QxQ; 13.BxQ, B-Kt2; 14.BxB, R-Q; 15. CastlesQRch, K-K2; 16.B-B6ch, KxQB; 17.RxR and comes out the exchange ahead. For black to move 11.B-K2 would result in the same continuation as in the text.

(f)—Something must be done to avoid the fatal discovered check.

(g)—If 13....QxRch, black loses as follows: 14.RxQ, BxB; 15.QxQBPch, Kt-B3 (if B-B3, 16.QxPch and wins the rook); 16.P-QKt4ch, K-B5; 17.Q-B7ch, K-Kt4 (KxP allows the Q to mate); 18.QxPch, etc.

(h)—If K-R4, 16.R-Q5ch followed by Q-Kt4 mate.

71. The record of a "correspondence" chess game; in such games, either of the two players has many hours to decide on his next (mailed) move.

The solution to Problem No. 75 (Tournament Problem No. 13), by Winston Kock, is:

1. Q-R7, KxKt; 2. Kt-Q3 mate.
———, QxQ; 2. Kt-B6 mate.
———, QxR; 2. Q-R1 mate.
———, Q other; 2. Q-R1 or KtB6 mate acc.
———, Kt any; 2. Kt-B3 mate.
———, RxKt(B4); 2. QxR mate.
———, RxKt(Q4); 2. KtxP mate.
———, R other; 2. KtxP mate.

This problem created havoc with the chances of many solvers for winning a prize in the solving contest. It suggests the old nursery rhyme about "Humpty Dumpty" having a great fall and also is strangely reminiscent of that old-time pathetic ballad, "Down Went McGinty." Anyhow, something or somebody fell mighty hard. Last week there were eighteen solvers tied for first place. Since then five of the leaders bungled this one and were cunningly misled by a very deceptive try which the foxy and astute Kock had prepared for the unwary. Several other solvers who were very close to the leaders also were deceived by this try. All in all, this problem "upset" no less than twelve of our solvers. Some record, that! Two of the contestants thought KtxQ was the correct key, but that move is defeated by Kt-K5! The other ten overlooked, when they played B-B6 for a key, the defense of black that keeps white from mating. The defense is Q-Q3 pinning the white knight against the king so that it can not move and give mate.

72. *Newspaper account of the difficulties experienced in solving a two-move chess problem.*

PROBLEM NO. 50
By WINSTON KOCK
Black—8 Pieces

White—7 Pieces

White to play and mate in three moves.
White—K on KR4, Q on QB6, B on QR1, Kt's on QR3 and QB4, P's on KB2 and KKt5.
Black—K on KB4, Kt's on KB2 and KR4, B on QKt8, P's on QR3, K3, KB5 and KKt3.

This week we have made a departure from our usual custom of presenting a two-mover for our solvers and dish up for their chess repast the above appetizing three-mover. As this is the fiftieth problem published in this column, it is quite appropriate that the composer is one of our local contingent and also one of our cleverest. Winston Kock, as our solvers are aware, has composed many baffling and intricate two-movers for their edification. The above problem, however, is his maiden three-mover.

At time of going to press not a single correct answer had been received to Problem No. 50, by Winston Kock, so we have decided to defer publishing the solution until next Saturday and will grant our solving corps until Tuesday, March 6, to submit solutions to this problem, which seems to have them baffled. Winston is as pleased at mystifying the solvers as the baby born with the proverbial silver spoon in its mouth or the boy who has received his first drum and naturally is heralding from the housetops about how good he is. Bless his little heart, let him enjoy the thrill that comes once in a lifetime!

We halfway expected that our solving corps would be puzzled when we submitted the problem to them, as a great majority are novices who have never tried solving anything but two movers. Three-move problems are much more difficult to solve, as they allow the composer more latitude to present his ideas and elaborate his ingeniousness. We believe that for a while it will be best for all concerned to continue to dish up two movers for our solving band.

We desire to say that Mr. Kock's problem has a legitimate solution and it can be solved. We also wish to state that P-B3 will not solve it.

We admire the courage of the many who tried but failed to find its clever solution and hope they will try again. One of our old-time solvers says it can't be done.

73. *A newspaper account of the first publication of a three-move chess problem, noting the difficulties experienced in solving it.*

74. Report of the author's appointment to a board of directors.

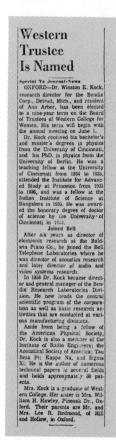

75. Report of the author's election to a board of trustees.

THIS IS TO CERTIFY THAT

WINSTON E. KOCK

WAS ORDAINED A **RULING ELDER**

in the _____ First _____ Presbyterian Church

_____ Ann Arbor _____ _____ Michigan _____
CITY STATE

of _____ Detroit _____ Presbytery.

according to the Constitution of The United Presbyterian Church in the United States of America

_____ January 24, 1960 _____
DATE

MODERATOR OF SESSION

CLERK OF SESSION

76. Ruling elder certificate.

UNIVERSITY OF CINCINNATI
CINCINNATI, OHIO
THE PRESIDENT

April 1, 1952

Dear Dr. Koch:

I am delighted to inform you that the Board of
Directors of the University of Cincinnati, at a
meeting today, voted to confer upon you the degree
of Doctor of Science at the 1952 Commencement
exercises on June 6. Action was taken by the Board
upon the unanimous recommendation of the University
Committee on Honorary Degrees.

This award of the University's honorary Sc. D.
is in recognition of your outstanding and sustained
record of achievement in important fields of science.

I am sure that, as a graduate of the University,
you will feel a special pride in this honor which
your Alma Mater is to bestow upon you.

The University does not confer honorary degrees
in absentia and accordingly your presence at the
exercises on Friday, June 6, will be necessary. I
trust that nothing will prevent your attendance.

For the present, the announcement of your and
three other degrees is confidential. The University
will make a public announcement in due time.

With heartiest congratulations and every good
personal wish, I am

Sincerely,

Raymond Walters

Raymond Walters

Dr. Winston Koch
Bell Telephone Laboratories
Murray Hill
New Jersey

RW:pb

77. Notification of award of an honorary doctorate.

78. *Award of Merit plaque.*

Eminent Member

WINSTON KOCK

Dr. Winston Kock, Vice President and Chief Scientist of Bendix Corporation was inducted into Eminent Membership on November 3, 1966 at Boston, Massachusetts. The induction and a luncheon that followed was held in connection with the program of NEREM. The induction team was made up of Clyde Hyde, National President; John A. Tucker, Past National Director, and Bruce D. Wedlock, Past President of Boston Alumni Chapter. The Program was in charge of

Ronald Goldman, President of Boston Alumni Chapter.

Dr. Kock recently returned to Bendix after having served as the Director of the Electronic Research Center of the National Aeronautics and Space Administration at Cambridge, Mass., since September, 1964. In this capacity, he guided the Center during the formative years of its $60 million facilities design and construction plans and pioneering work in space and aeronautical electronics. Under Dr. Kock's

direction an initial cadre of 70 scientists and engineers was developed to a staff of 550.

Dr. Kock first joined Bendix in 1956 as Chief Scientist of its Systems Division. He became Director and General Manager of the Research Laboratories Division in 1958 and was made Vice President of Research for the corporation in 1962. He served in that position until he joined NASA.

Before joining Bendix in 1956, Dr. Kock was associated with Bell Telephone Laboratories as Director of Acoustics Research.

He took part in the early stages of the development of the "picturephone." During World War II, research in the field of microwave antennas led to his invention of the microwave lens antenna used extensively in radar and transcontinental microwave relay systems for transmission of television programs. He developed the Baldwin electronic organ when he served as Director of Electronic Research for the Baldwin Piano Company from 1936 to 1942.

Dr. Kock received his electrical engineering and master of science degrees from the University of Cincinnati and Ph.D. in physics from the University of Berlin in 1934. He received an honorary doctor of science degree from the University of Cincinnati in 1952.

He has been consultant to the Secretary of Defense, Department of the Army, Office of Naval Research, National Academy of Sciences and the Department of Defense Advisory Panel on Electronics. He is a Fellow of the American Physical Society, Acoustical Society of America, Institute of Electronic and Electrical Engineers, and a member of the American Institute of Physics, National Security Industrial Association and Industrial Research Institute.

OUR COVER

Dr. Winston E. Kock is the first in history to win the Eta Kappa Nu Grand Slam, i.e., all of the honors conferred by the Association. He was elected a regular member by Tau chapter in his college days and later was presented their Award of Merit for Outstanding Alumni. He was selected the Outstanding Young Electrical Engineer in the United States in 1938. In 1944 he was elected National President. Finally in 1966 he was elected Eminent Member. BRIDGE sends its congratulations and best wishes to Dr. Kock.

79. Eminent Member award.

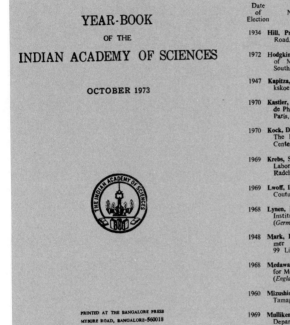

YEAR-BOOK

OF THE

INDIAN ACADEMY OF SCIENCES

OCTOBER 1973

PRINTED AT THE BANGALORE PRESS
MYSORE ROAD, BANGALORE-560018

32

Date of Election	Name, Designation and Address
1934	**Hill, Professor A. V.,** Nobel Laureate, 11A, Chaucer Road, Cambridge (*England*)
1972	**Hodgkin Psofessor D. C.** Nobel Laureate, Laboratory of Molecular Biophysics, Department of Zoology South Parks Road, Oxford OX1 3PS. (*England*)
1947	**Kapitza, Professor P.,** Director, Physical Institute, Kaluzkskoe Schosse 32, Moscow (*S.S.R.*)
1970	**Kastler, Professor Alfred,** Nobel Laureate, Laboratoire de Physique, Ecole Normale Superieure, Université de Paris, 24, Rue Lhomond, Paris (5e) (*France*)
1970	**Kock, Dr. Winston E.,** Vice-President and Chief Scientist, The Bendix Corporation, Executive Offices, Bendix Center, Southfield, Michigan 48075 (*U.S.A.*)
1969	**Krebs, Sir H. A.,** Nobel Laureate, Metabolic Research Laboratory, Nutfield Department of Clinical Medicine, Radcliffe Infirmary, Oxford (*England*)
1969	**Lwoff, Dr. Andre,** Nobel Laureate, 16, Avenue Vaillant-Couturier, 94, Villejuif, Boite Postale No. 8 (*France*)
1968	**Lynen, Professor Dr. F.,** Nobel Laureae, Max-Planckt Institut für Zellchemie, 8 Munchen 2, Karlstrasse 23 (*Germany*)
1948	**Mark, Professor Hermann,** Director, Institute of Polymer Research, Polytechnic Institute of Brooklyn 99 Livingston Street, Brooklyn 2, New York (*U.S.A.*)
1968	**Medawar, Sir Peter,** Nobel Laureate, National Institute for Medical Research, Mill Hill 3666, London, N.W, 7 (*England*)
1960	**Mizushima, Professor Dr. San-ichiro,** 698 2-Chome, Tamagawa-Gegenchofa, Setagya-Ku, Tokyo (*Japan*)
1969	**Mulliken, Professor Roberts,** Nobel Laureate, LMSS— Department of Physics, The University of Chicago, 1100 East 58th Street, Chicago, Illinois 60637 (*U.S.A.*)

80. *Indian Academy of Sciences Fellow award.*

```
                        THE INSTITUTE FOR ADVANCED STUDY
Director                          1935-6
     Dr. Abraham Flexner (M)
     Office: 20 Nassau St., telephone 497;   Home: The Princeton Inn, telephone 1020
                           School of Mathematics
                                Fine Hall
Prof. James W. Alexander (M), 29 Cleveland Lane - telephone 32
Dr. Reinhold Baer
Prof. Paul Bernays (S), 87 Jefferson Road - telephone 1035-M
Dr. Willard E. Bleick (S), 42 Wiggins Street - telephone 1749-W
Dr. Leonard M. Blumenthal (M), 28 Vandeventer Avenue - telephone 542-J
Dr. Louis P. Bouckaert (S), Graduate College Room 181 - telephone 300 G.C.Entry 18
Prof. Gregory Breit (M), 69 Patton Avenue - telephone 1375
Prof. Leonard Carlitz (M), 345 Nassau Street - no telephone
Dr. John F. Carlson (S), 159 Nassau Street - telephone 57-W
Prof. Eduard Čech (M*), 2 LeVake Place - telephone 201-R
Prof. Edward W. Chittenden (M*), 107 Moore Street - telephone 130-W
Dr. James A. Clarkson (M), 14 Spruce Street - no telephone
Dr. Alfred H. Clifford (S), 43 Vandeventer Avenue - telephone 373-W
Dr. George Comenetz (S), 19 Vandeventer Avenue - telephone 160-J
Dr. Edward H. Cutler (S), 9 Madison Street - telephone 420-J
Prof. Arnold Dresden (M*), Nassau Club - telephone 580
Prof. Albert Einstein (M), 112 Mercer Street - telephone
Prof. Philip Franklin (M), 94 Bayard Lane - telephone 714-J
Prof. Bennington P. Gill (S**), 54 Murray Place - telephone 1344-M
Mr. James W. Givens, Jr. (S), 43 Vandeventer Avenue - telephone 373-W
Dr. Kurt Goedel (S), 23 Madison Street - telephone 654
Prof. Lawrence M. Graves (M), 17 Aiken Avenue - telephone 894-W
Dr. Banesh Hoffmann (S), 13 Park Place - telephone 201-M
Dr. Spofford H. Kimball (M*), 13 Park Place - telephone 201-M
Dr. Winston E. Kock
Prof. Max von Laue (M*), Nassau Club - telephone 580
Dr. Norman Levinson (S), 49 Wiggins Street - telephone 555-R
Dr. Robert S. Martin (S), 19 Vandeventer Avenue - telephone 160-J
Dr. William T. Martin (S), 65 Wiggins Street - telephone 1349-M
Prof. Walther Mayer (M), 26 Linden Lane - no telephone
Prof. Marston Morse (S), Graduate College Room 1414 - telephone 1523-W
Dr. Francis J. Murray (M), 43 Linden Lane - no telephone
Dr. Sumner B. Myers (S), 42 Wiggins Street - telephone 1749-W
Prof. John von Neumann (M), Corner Elm Road and Cleveland Lane - telephone 1862
Prof. Wolfgang Pauli (M), 56 Princeton Avenue - telephone 1294
Dr. Arthur E. Pitcher (S), 98 Jefferson Road - telephone 171
Mr. Maurice H. L. Pryce (S), Grad. College Room 162 - telephone 300 G.C.Entry 16
Dr. William C. Randels (S), 42 Wiggins Street - telephone 1749-W
Mr. Louis N. Ridenour, Jr. (M), 12 Princeton Avenue - telephone
Dr. Morris E. Rose (M), 193 Van Nostrand Avenue, Jersey City, N.J. - no telephone
Dr. Nathan Rosen (M), 14 Spruce Street - no telephone
Dr. Otto F. G. Schilling (S), 8 Morven Place - telephone 667
Dr. Roman Smoluchowski (S), 24 Dickinson Street - telephone 334-W
Dr. Martin H. Stobbe (S), 11 Park Place - telephone 914-J
Mr. Eric D. Tagg (S), Graduate College Room 182 - telephone 300 G.C.Entry 18
Dr. Abraham H. Taub (M), 46 Park Place - telephone 926-R
Dr. Stanislaw M. Ulam
Dr. Gaston Van der Lyn (S), Grad. College Room 178 - telephone 300 G.C.Entry 17
Prof. Oswald Veblen (M), 58 Battle Road - telephone 958
Prof. Hermann Weyl (M), 220 Mercer Street - telephone 2171
Dr. Lee R. Wilcox (S), 43 Vandeventer Avenue - telephone 373-W
Mr. Shaun Wylie (S), Graduate College Room 193 - telephone 300 G.C.Entry 19
Dr. Leo Zippin (M), 217 Nassau Street - telephone 1214-W
M - married                              * Wife not in Princeton, October, 1935
S - single                              ** Mother in Princeton (Mrs. Joseph B. Gill)
```

81. Member list at the Institute for Advanced Study.

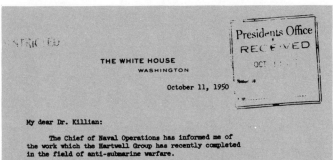

THE WHITE HOUSE
WASHINGTON

October 11, 1950

President's Office
RECEIVED
OCT

My dear Dr. Killian:

The Chief of Naval Operations has informed me of the work which the Hartwell Group has recently completed in the field of anti-submarine warfare.

I wish to take this opportunity of expressing to you and through you to all of the members of the Hartwell Group not only my very great interest in your undertaking, but also my appreciation of your fine accomplishments. As you are aware, many of the recommendations made by the Group can and will be placed in effect by the Department of the Navy alone. Such recommendations as involve other Departments will be coordinated with those Departments.

It is my hope that the individual members of the Hartwell Group will not consider that the submission of their Report completes their interest in the important project of helping to build the best possible anti-submarine structure for the Navy. The contribution made by the Group has been impressive and will, I am sure, materially increase our strength. Continued interest by all those concerned will further improve matters.

With kindest regards and, again, my deep appreciation of the Group's splendid accomplishment.

Very sincerely,

Harry Truman

Dr. J. R. Killian, Jr.
President, Massachusetts Institute of Technology
Cambridge, Massachusetts

82. A letter from President Truman to MIT President Killian expressing his appreciation of the Hartwell Group's accomplishments.

Reichsminister a.D. Freiherr von Braun
und Freifrau von Braun

bitten Mr. Winston E. Kock am Sonnabend den 3.März 1934
von 8 Uhr ab den Abend bei ihnen im kleinen Kreise ver-
brimgen zu wollen.

Potsdamer Privatstrasse 121 c Tel. Kurfürst B 1 418o

U.A.w.g.   Bitte Smoking

*83. Invitation from Baron and Baroness von Braun, parents of Wernher von Braun.
("Bitte Smoking" means "Please wear tuxedo.") The author's exchange fellowship
(Cincinnati to Berlin) involved, as the other person, Siegesmund von Braun (Berlin to
the University of Cincinnati), Wernher von Braun's older brother.*

*84. A group photo taken at the author's Ann Arbor home. Left to right: Nobel prize
winners Peter Debye and John Bardeen, the author, Nobel Prize winner William
Shockley, and Ralph Sawyer, Vice-President for Research at the University of
Michigan.*

W. E. Kock   Murray Hill

That I.R.E. paper was a mighty
nice presentation of a splendid contribution
on your part — the electromagnetic lens.

I couldn't help but reflect upon
the great contribution to microwave
radio that has come out of the Bell
System in the combination of
        wave guide & lens antenna,
Congratulations to you and Southworth
& Company!
                        Lloyd Espenschied

85. *A letter from the inventor of the coaxial cable.*

86. Zino Francescatti playing a violin selection with author standing behind him.

87. Innovators relaxing at the home of Dr. and Mrs. John M. Ide in Washington, D.C.

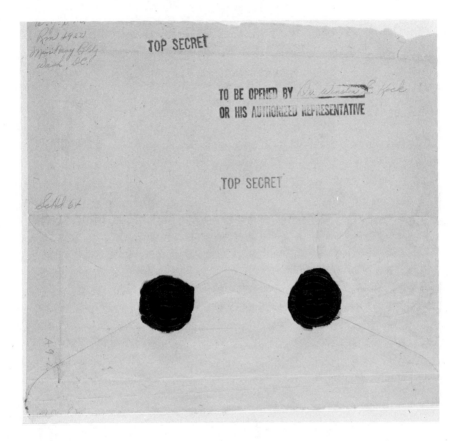

88. An envelope for a top secret document.

89. As Director of the NASA Center, the author is "induced" to take part in the celebration of a motion picture premiere in Boston.

DEPARTMENT OF COMMERCE
UNITED STATES PATENT OFFICE
WASHINGTON

DEC 29

Serial Number 417,389   Filed   October 31, 1941

For   Oscillation Generator

Applicant Winston E. Kock

Assignee The Baldwin Company

NOTICE

To the applicant above named, his heirs, and any and
all his assignees, attorneys and agents:
Under the provisions of the Act of October 6, 1917
(Public No. 80) as amended July 1, 1940 (Public No. 700)
as amended August 21, 1941 (Public No. 239), you are hereby
notified that your application as above identified has been
found to contain subject matter disclosure of which might be
detrimental to the public safety or defense, and you are
hereby ordered to in nowise publish or disclose the invention or disclosure of said application, but to keep the same
secret (except by written permission first obtained of the
Commissioner of Patents), under the penalties of the amended
Act. This application must be prosecuted under the Rules of
Practice until a notice is received from the office that all
the claims then in the case are allowable. Such notice closes
the prosecution of the case. Furthermore, if previously allowed and now withdrawn from issue the prosecution of the case
is likewise closed. When the application is in condition for
allowance it will be withheld from issue during such period
or periods as the national interest requires.
This order should not be construed in any way to mean
that the Government has adopted or contemplates adoption of
the alleged invention disclosed in this application, nor is
it any indication of the value of such invention. It is
recommended that you tender this invention to the Government
of the United States. Tender may be effected by communicating
directly with the Secretary of War or the Secretary of the
Navy.

Conway P. Coe
                                          Commissioner.

90. The author's first "secrecy notice."

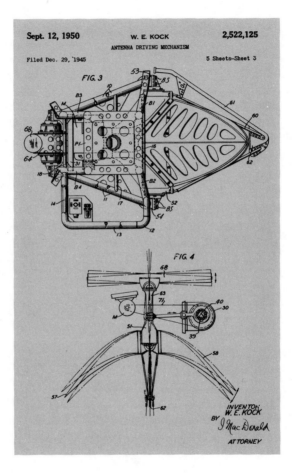

Sept. 12, 1950   W. E. KOCK   2,522,125

ANTENNA DRIVING MECHANISM

Filed Dec. 29, 1945   5 Sheets—Sheet 3

91. A page from the patent on the Mark 13 radar antenna.

ADDRESS REPLY TO
BUREAU OF ORDNANCE, NAVY DEPARTMENT
AND REFER TO

NAVY DEPARTMENT
BUREAU OF ORDNANCE
WASHINGTON 25, D. C.

15 June 1946

Subject: Naval Ordnance Development Award.

Enclosure: (A) Certificate for Exceptional Service to
                Naval Ordnance Development.

Dear Mr. Kock:

        It is the great pleasure of the Chief of
the Bureau of Ordnance to confer upon you the Naval Ordnance Development Award which has been granted in recognition of your exceptional service to the research and
development of naval ordnance.

        The congratulations of the Bureau of Ord-
nance are extended to you for your outstanding performance
in connection with the research and development of electrical design of the "rocking" type antenna used in the Mark 13
radar.

        The Certificate for Exceptional Service to
Naval Ordnance Development and the lapel emblem are the
symbol of appreciation from the Bureau of Ordnance and from
the entire Navy for your unrelaxed efforts and keen technical proficiency which you have consistently displayed.

                   Very truly yours,

                   G. F. HUSSEY, JR.
                Vice Admiral, U. S. Navy
            Chief of the Bureau of Ordnance

Mr. W. E. Kock
Bell Telephone Laboratories, Inc.
463 West Street
New York, New York

92. The letter from Admiral Hussey praising the performance of the Mark 13 radar.

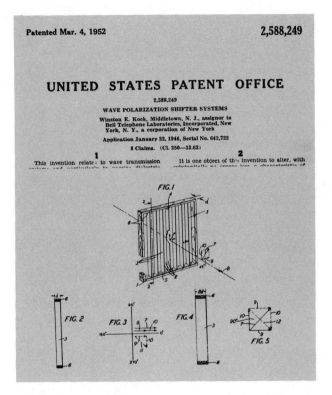

93. *A page from the author's patent describing the polarization rotating device.*

screens on any one vehicle will be uniformly oriented at 45° from the vertical. The driver can then see ahead in the light of his own headlamps, but his windshield screen will quench the direct light from the headlamps of a car coming in the opposite direction.

The photographs in Fig. 9 below will serve to illustrate the effect of this procedure in practice. The headlights of the automobile were covered

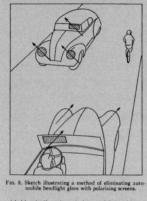

FIG. 8. Sketch illustrating a method of eliminating automobile headlight glare with polarizing screens.

with identically oriented sheets of Polaroid and pointed at the camera. At the same time, the automobile was illuminated by the headlights of another motor car standing to the right of the camera. For the upper picture, taken without a polarizing filter over the camera lens, the time of exposure was necessarily very short. For the lower picture, the camera lens was covered with a Polaroid filter oriented to quench the direct headlight; and the time of exposure was about five hundred times as long as the other. In both cases, a minus-blue filter was used to compensate for the disparity between the blue sensitivity of the eye and of the photographic film. Much of the apparent brightness of the headlights in the

lower picture is due to light reflected from the outer surfaces.

However, this photograph can hardly convey the astonishing impression created by being able to read the number plate between the headlights and to see the occupants of an automobile approaching head-on. At the same time, the illumination of the road ahead remains as effective as it was before the approaching car came into view.

PHOTOMETRICAL USES IN OPTICAL INSTRUMENTS

When one of two superposed sheets of Polaroid is rotated, the relative intensity of visible light transmitted by the ensemble follows the familiar cosine-squared law very closely.[10] The error is relatively very small except within a few degrees of the extinction setting where the effect of the slight defect of polarization becomes relatively large. Such a combination yields a range of relative intensities of at least two hundred to one; and this range can be extended enormously by

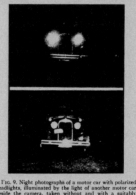

FIG. 9. Night photographs of a motor car with polarized headlights, illuminated by the light of another motor car beside the camera, taken without and with a suitably oriented polarizing filter over the camera lens.

94. *A page from a 1938 publication using a linear polarizing material (Polaroid) to reduce the glare from oncoming car headlights. Circular polarization material has become available recently. It will also stop headlight glare because the light from the oncoming car is automatically in the wrong sense of rotation to pass through the windshield polarizing screen. (It also works perfectly when both cars are not exactly vertical; the linear material loses it maximum effectiveness in such a situation.) Also, circular polarization can improve visibility during fog or mist.*

OCTOBER, 1937   J. A. S. A.   VOLUME 9

A New Interpretation of the Results of Experiments on the Differential Pitch Sensitivity of the Ear

Winston E. Kock*
The Institute for Advanced Study, Princeton, New Jersey
(Received May 19, 1937)

Introduction

OF the many investigations on the differential pitch sensitivity of the ear (the minimum change in frequency which the ear is able to detect at any one frequency and sensation level), probably the two most important are those of Knudsen[1] and Shower and Biddulph.[2] However, certain considerations arising from the application of the principle of uncertainty to sound[3] show that, in both of these cases, the results would have been very much the same even if a perfect receiver, i.e., one which could detect an infinitesimal change in frequency, had been employed in place of the ear. In other words, the experimental methods employed introduced undesired effects which masked the true differential pitch sensitivity of the ear.

It is the purpose of this paper to point out how many of the results of these two investigations could have been quantitatively predicted from theoretical considerations, and to show how deviations of the actual results from predicted results can be used to estimate the actual differential pitch sensitivity in certain cases. Since these two papers have been taken as authoritative by recent books on sound,[4] a critical consideration of the results would appear to be of value.

Preliminary Considerations

The principle of uncertainty $\Delta p \Delta q \approx h$ was first applied by Heisenberg to elementary particles; it states that uncertainty of the position (q) times the uncertainty of momentum (p) of a particle is *of the order of* Planck's constant h. When this principle is put in the form[5] $\Delta E \Delta t \approx h$

and applied to photons (so that $E = h\nu$) we obtain the relation $\Delta \nu \Delta t \approx 1$. Applying this to the wave train of the photon, we note that a wave train lasting for the time Δt has the uncertainty of frequency $\Delta \nu$. However, this property is intrinsic in any type of wave motion. Thus, although a wave train of infinite length is composed of but one single frequency, a train of finite length is not, for it can be represented by a band of infinitely long trains of Fourier waves. The shorter the duration of the vibration, the wider the frequency band covered by it. Furthermore, it is known that in the curve which gives intensity as a function of the frequency, the distance between two frequencies for which the intensity has half its maximum value is given by[6] $\Delta \nu = 1/\Delta t$, i.e., $\Delta \nu \Delta t = 1$.

If the wave train just referred to happened to be a sound wave, it would likewise possess such a frequency distribution, and the shorter the duration of the wave train, the more difficult it would be for any receiver, including the ear, to determine its exact frequency because of this frequency "spread." We observe that for two notes having the same duration, the higher pitched one can be more accurately ascertained for $\Delta \nu/\nu$ is smaller. An experimental verification of this fact can be found in a recent paper.[7] Similarly a note rich in harmonics permits of more exact determination because the higher frequency harmonics possess less frequency spread. Increasing the intensity of a pure tone increases the subjective harmonic content and hence the accuracy of perception.

Let us turn now to a consideration of a wave of infinite length whereby, however, the frequency is periodically altered. This would correspond to a frequency vibrato if the frequency were varied continuously back and forth, or to certain types

* Now with the Baldwin Piano Company, Cincinnati, O.
[1] V. O. Knudsen, Phys. Rev. 21, 84 (1923).
[2] E. G. Shower and R. Biddulph, J. Acous. Soc. Am. 3, 275 (1931).
[3] W. E. Kock, J. Acous. Soc. Am. 7, 56 (1935).
[4] E.g., Mills, *A Fugue in Cycles and Bels* and *Handbuch der Physik*, Vol. 8 (J. Springer), p. 807.
[5] This relation holds for any pair of canonically conjugate variables; it must therefore hold for H and t, for from Hamilton's equation $dq/dt = \partial H/\partial p$, it is seen that for $H = E \rightarrow p$, $dq/dt = 1$, i.e., $q \rightarrow t$.
[6] Joos, *Theoretical Physics*, Blackie and Son, 1934, p. 657.
[7] Bürck, Kotowski, and Lichte, Ann. d. Physik 25, 433 (1936). In this paper it will be observed that above 2000 cycles the accuracy of perception does not continue to increase with increased frequency. This is due to the decreasing sensitivity of the ear above 2000 cycles.

95. *The first page of a 1937 paper, the conclusions of which were later corroborated by Dennis Gabor, the inventor of holography. Figs. 1 and 3 of this paper were used in* Psychoacoustics, *by J. Donald Harris (Bobbs-Merrill, 1972).*

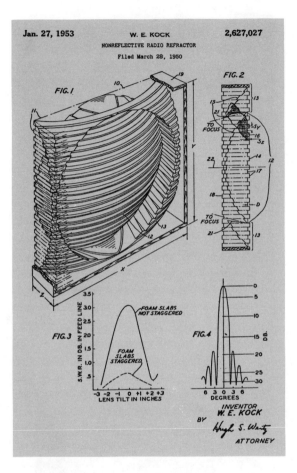

Jan. 27, 1953   W. E. KOCK   2,627,027
                        NONREFLECTIVE RADIO REFRACTOR
                           Filed March 28, 1950

96. *Phase quadrature "coating" on a microwave lens.*

97. NASA's Lowell Rosen (right) shows holography inventor Dennis Gabor (center) a reconstructed hologram while the author looks on. Courtesy NASA.

98. The author (left) and George Stroke, the pioneer in making laser holograms who first proposed the term "holography" (taken at the 1973 NATO Advanced Study Seminar in Capri, Italy).

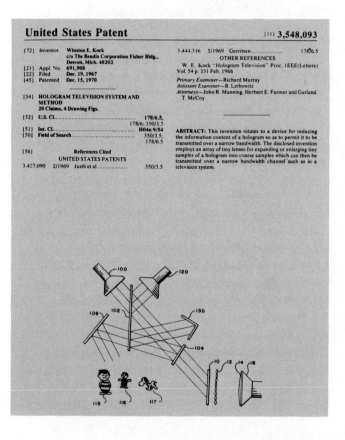

United States Patent [11] 3,548,093

[72] Inventor **Winston E. Kock**
c/o The Bendix Corporation Fisher Bldg.,
Detroit, Mich. 48202
[21] Appl. No. 691,908
[22] Filed Dec. 19, 1967
[45] Patented Dec. 15, 1970

[54] HOLOGRAM TELEVISION SYSTEM AND METHOD
20 Claims, 4 Drawing Figs.

[52] U.S. Cl. 178/6.5,
178/6; 350/3.5
[51] Int. Cl. H04n 9/54
[50] Field of Search 350/3.5;
178/6.5

[56] **References Cited**
UNITED STATES PATENTS
3,427,090 2/1969 Justh et al. 350/3.5

3,444,316 5/1969 Gerritsen................. 178/6.5
OTHER REFERENCES
W. E. Kock "Hologram Television" Proc. IEEE(Letters) Vol. 54 p. 331 Feb. 1966

Primary Examiner—Richard Murray
Assistant Examiner—B. Leibowitz
Attorneys—John R. Manning, Herbert E. Farmer and Garland T. McCoy

ABSTRACT: This invention relates to a device for reducing the information content of a hologram so as to permit it to be transmitted over a narrow bandwidth. The disclosed invention employs an array of tiny lenses for expanding or enlarging tiny samples of a hologram into coarse samples which can then be transmitted over a narrow bandwidth channel such as in a television system.

99. First page of a patent describing a technique to make holography amenable to transmission over standard television channels.

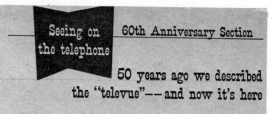

Seeing on the telephone | 60th Anniversary Section

50 years ago we described the "televue"—— and now it's here

It is now possible to *see* as well as to *hear* on the telephone.

You can even carry on a conversation with someone, and then hang up and pull out of the receiver a picture of the person who was on the other end of the line.

All this is possible, although right now it is actually being done only by engineers of the Bell Telephone Laboratories. They have carried on talk-and-look exchanges over distances as long as those separating New York from Los Angeles. The apparatus has been described in a technical paper before The Institute of Radio Engineers. Bell Labs officials caution that the device is "still undergoing development and evalua-

1906 This sketch of a sales-girl showing a hat accompanied our early description of the "televue".

1956 One of the phone company's test models. Compare camera "eye" above screen to the 1906 sketch.

tion and is not ready for manufacture or commercial use."

Eventually, they expect, refinements of the picture-phone will "overcome the cost barrier."

A big step toward commercial feasibility was taken when the engineers succeeded in transmitting the pictures over one pair of wires identical with the pair that now carry voices to millions of phones. There remains the problem of which of three types of receiver is best for the lowest price.

All three types show head and shoulders of the caller in the picture, with "facial expressions readily apparent" but in considerably less detail than a TV picture. All have a mirror-like frame to help each caller keep himself in focus one to two feet away from the phone. All have picture screens either one by one and a half inches or two by three inches in size. All operate optionally and separately from the voice receiver, and can be turned on or off by either party before or during a conversation without affecting the sound.

All transmit a clear likeness in natural daylight or in the ordinary brightness of a lighted room.

On one of the three types of apparatus, the picture is developed camera-fashion on a lighted film in the receiver. It is this print which can be removed from the picture-phone after a call has been finished.

The nation's press, in reporting the Bell Labs' announcement of the picture-phone in August, 1956, noted that Bell engineers have worked on this invention since 1927. It is no reflection on their accomplishment to record the fact that the story of a seeing telephone goes considerably further back than that.

The American Weekly printed the following on June 3, 1906, in reporting on a "seeing telephone" invented by J. B. Fowler of Portland, Oregon, and tested over a distance of more than 6,000 feet:

"The televue will revolutionize the conditions of modern life, perhaps even more completely than the telegraph or the telephone. . . . With the simple telephone it is difficult to buy goods with confidence because naturally one wishes to see them before buying.

"But this difficulty will be removed by the televue. It will simplify the task of the housewife enormously. She will be able to buy dress goods and provisions and do all her shopping by televue. The salesgirl will hold up the article desired before the televue transmitter and say: 'How will this hat suit Madame?' " ◄◄

100. A 1906 sketch of a picture telephone.

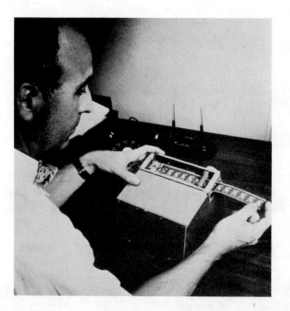

101. R. L. Miller and the permanent record Picturephone device.

102. An early newspaper report on the Picturephone.

103. *Electronic Daily features the Picturephone story on its cover.*

Manuscript Service   *Texts of Broadcasts*

Radio Reports, Inc.

220 EAST 42nd STREET
NEW YORK 17, N. Y.
MUrray Hill 7-6658

Special for PACIFIC TELEPHONE AND TELEGRAPH COMPANY

August 23, 1956

BELL TELEPHONE ENGINEER DISCUSSES NEW PICTURE TELEPHONE

<u>Eyewitness</u> at 7:15 P.M. over KTLA-TV (Los Angeles):

> KEN GRAUE AND MR. KOCK WERE SEEN IN MEDIUM
> CLOSEUP. A STILL PHOTOGRAPH OF THE NEW
> PICTURE TELEPHONE WAS SHOWN IN CLOSEUP.
> MEDIUM CLOSEUP WAS AGAIN SHOWN OF MR. KOCK
> AS HE DESCRIBED THE COST OF THE NEW TELEPHONE.

GRAUE: "Eyewitness to the development of the new picture phone is Winston Kock, BELL LABORATORIES engineer, from Murray Hill, New York. He's here in Los Angeles attending the Western Electronics Show and Convention. In just a moment, Mr. Kock, we're going to have the opportunity of seeing via film the operation of the picture telephone, and perhaps before we do you can answer just a few questions. Why until recently has the picture telephone been impossible?"

KOCK: "Well, it's required very expensive coaxial cables to transmit television pictures before, and by this new technique where we record the picture and send it out--we can use the normal telephone lines instead."

GRAUE: "Well, then, actually there wouldn't be any installation difficulties as far as putting such a phone into the home, is that correct, Mr. Kock?"

KOCK: "The installation would be very, very simple, when it is developed to a commercial point, which it is not as yet."

GRAUE: "You brought along a picture of the picture phone that we're holding here, and perhaps I can point out just a couple of things and I know that you can see them and tell us what they are on our monitor screen. It looks to me like that you have a lens directly at the top of the box. Is that correct?"

————CONFIDENTIAL————

Chicago • Cleveland • Detroit • Los Angeles • New England • New York • Philadelphia • San Francisco • Washington

104. First page of the KTLA-TV television interview.

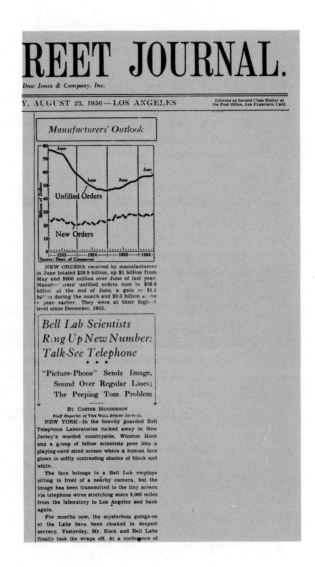

REET JOURNAL.

Dow Jones & Company. Inc.

Y. AUGUST 23, 1956 — LOS ANGELES

Entered as Second Class Matter at
the Post Office, San Francisco, Calif.

Manufacturers' Outlook

Unfilled Orders

New Orders

Source: Dept. of Commerce

NEW ORDERS received by manufacturers
in June totaled $28.9 billion, up $1 billion from
May and $600 million over June of last year.
Manufacturers' unfilled orders rose to $58.6
billion at the end of June, a gain of $1.1
billion during the month and $9.3 billion above
a year earlier. They were at their highest
level since December, 1953.

Bell Lab Scientists
Ring Up New Number:
Talk-See Telephone
• • •

"Picture-Phone" Sends Image,
Sound Over Regular Lines;
The Peeping Tom Problem

BY CARTER HENDERSON
Staff Reporter of THE WALL STREET JOURNAL

NEW YORK—In the heavily guarded Bell
Telephone Laboratories tucked away in New
Jersey's wooded countryside, Winston Kock
and a group of fellow scientists peer into a
playing-card sized screen where a human face
glows in softly contrasting shades of black and
white.

The face belongs to a Bell Lab employe
sitting in front of a nearby camera, but the
image has been transmitted to the tiny screen
via telephone wires stretching some 6,000 miles
from the laboratory to Los Angeles and back
again.

For months now, the mysterious goings-on
at the Labs have been cloaked in deepest
secrecy. Yesterday, Mr. Kock and Bell Labs
finally took the wraps off. At a conference of

105. Front-page portion of the Wall Street Journal coverage.

LOS ANGELES EVENING
HERALD EXPRESS

PRESS
CLIPPINGS

Date ___August 23, 1956___

Edition ___Latest News___ Page1 Sec.D

Just Dial and See Them

Picture-Phone Seen Near

ONE EXPERIMENTAL MODEL OF PICTURE PHONE WITH 2x3 INCH PHOTO
Displayed by W. E. Kock of Bell Telephone Laboratories; Arrow Points to Camera
Lens; Knobs Are for Turning On the Set and Defining the Picture

It won't be long until you see a picture of the person at the other end of your telephone conversation, Bell Telephone Laboratories announced here today.

A technical description of the "picture-phone" system—the first to use ordinary telephone wires for transmission—was given last night at a joint meeting of the Institute of Radio Engineers and the West Coast Electronic Manufacturers' Association in convention here.

Telephone scientists have transmitted recognizable pictures, as far as from New York to Los Angeles. The experimental pictures vary in size from one by one and a half inches to two by three inches.

Unlike television, a new picture is displayed every two seconds, depicting head and shoulders and clear facial features.

The picture-phone promises to be commercially feasible because it's the first system of its kind to use a pair of ordinary telephone wires.

Only one other line, consisting of wires similar to telephone lines, would be installed on a telephone subscriber's premises, the convention was told.

It will be possible for a picture to be dialed, provided the switch on the picture equipment is turned on at both ends of the line.

The equipment which flashes black and white pictures, is still undergoing development and is not ready for manufacture or commercial use, it was pointed out.

106. The Los Angeles Herald Express story.

Montag, 27. August 1956

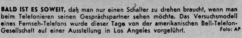

BALD IST ES SOWEIT, daß man nur einen Schalter zu drehen braucht, wenn man beim Telefonieren seinen Gesprächspartner sehen möchte. Das Versuchsmodell eines Fernseh-Telefons wurde dieser Tage von der amerikanischen Bell-Telefon-Gesellschaft auf einer Ausstellung in Los Angeles vorgeführt. Foto: AP

107. A German press article (the article and the "doctoring up" courtesy of Dr. Hermann Oberst).

AUTHOR INDEX

SUBJECT INDEX